Data Converters

Data Converters

G. B. Clayton

*Department of Physics,
Liverpool Polytechnic*

© G. B. Clayton 1982

All rights reserved. No part of this publication may be reproduced or transmitted, in any form or by any means, without permission.

First published 1982 by
THE MACMILLAN PRESS LTD
London and Basingstoke
Companies and representatives
throughout the world

Typeset by Reproduction Drawings Ltd
Sutton, Surrey

Printed in Hong Kong

ISBN 0 333 29494 7
ISBN 0 333 29495 5 pbk

The paperback edition of the book is sold subject to the condition that it shall not, by way of trade or otherwise, be lent, resold, hired out, or otherwise circulated without the publisher's prior consent in any form of binding or cover other than that in which it is published and without a similar condition including this condition being imposed on the subsequent purchaser.

Contents

Preface		viii
1	**Analog to Digital and Digital to Analog Conversion**	1
1.1	Usefulness of Converters	1
1.2	Digital Codes	2
1.3	Conversion Relationships	4
1.4	Decimal Codes	6
1.5	Bipolar Codes	8
1.6	Self-assessment Exercises	11
2	**Digital to Analog Converters**	14
2.1	Implementing a DAC Function	14
2.2	Analog Read-out	15
2.3	Use of R–2R Ladder Networks	16
2.4	Practical DACs	20
2.5	Examining the Performance of a DAC	21
2.6	Bipolar Coded DACs	26
2.7	Adding a Digital Data Storage Register to a DAC	32
2.8	Settling Time, Transient Errors, Glitches	34
2.9	Self-assessment Exercises	36
3	**Analog to Digital Conversion Techniques**	40
3.1	Feedback A/D Converters	40
3.2	Ramp-type A/D Converter	41
3.3	Tracking A/D Converter	43
3.4	Successive Approximation A/D Converter	49
3.5	Practical Successive Approximation A/D Converter Systems	52
3.6	Conversion Time of a Successive Approximation A/D Converter	59
3.7	Adding an Output Digital Storage Register to a Successive Approximation Converter	61

3.8	Integrating Techniques for A/D Conversion	62
3.9	Dual-slope A/D Conversion	63
3.10	Quantised-feedback A/D Conversion	67
3.11	Parallel A/D Conversion	72
3.12	Ratiometric A/D Conversion	73
3.13	Comparison of A/D Conversion Techniques	77
3.14	Self-assessment Exercises	79

4 Digital Analog Converter Applications — 82
4.1	Digitally Programmed Voltage and Current Sources	82
4.2	Digital Gain Control	87
4.3	Arithmetic Operations with Digital and Analog Variables	92
4.4	Digital Scale Setting	106
4.5	Digital Control of Frequency	107
4.6	Digital Control of a Linear Ramp	112
4.7	Generation of Functional Relationships	114
4.8	Digital Generation of Analog Waveforms	115
4.9	Self-assessment Exercises	119

5 Digital Processing of Analog Signals — 121
5.1	The Sampling Theorem	121
5.2	Sampling and Digitisation; Quantisation Noise	125
5.3	The Practical Implementation of Sampling and Digitisation	128
5.4	Sampling and Digitisation with a Successive Approximation A/D Converter	132
5.5	Non-linear Encoding	134
5.6	Applications of Sampled and Digitised Analog Signals	143
5.7	Converters in Data Acquisition Systems	152
5.8	Data Acquisition System Characterisation	158
5.9	Interfacing Converters to Microprocessors	164
5.10	Self-assessment Exercises	172

6 Practical Considerations — 176
6.1	Design Procedures	177
6.2	Application Objectives	177
6.3	Device Selection	179
6.4	Understanding Converter Performance Specifications	181
6.5	Ideal Conversion Functions	182
6.6	Error Specifications	186
6.7	Dynamic Response Parameters	197
6.8	Accuracy	197
6.9	Applying Converters	212
6.10	Self-assessment Exercises	218

References 220

Answers to Exercises 222

Index 239

Preface

The current upsurge in the use of digital processing techniques, brought about by developments in very large scale integration (VLSI), has forced the role of the interface elements between analog and digital systems to the forefront of attention. Data converters are these interface elements. Although they were once regarded by electronics engineers as expensive, rather specialised pieces of equipment, they are now commonplace. They are available as self-contained functional units, in both modular and low cost integrated circuit form. No one with an interest in modern electronic instrumentation techniques can afford to be without a sound knowledge and understanding of the operation, capabilities and limitations of these devices. This book sets out to provide a framework for such an understanding.

The first three chapters of the book discuss the principles underlying the operation of data converters. Stress is laid on those principles which the proliferation of available devices all have in common. Specific devices with pin connections are given, but only as a means of illustrating principles and to provide the reader with a convenient means of conducting his own experimentation, as an aid to the learning process. Experimental investigation of a converter device gives the engineer a practical understanding of device operation and does much to remove the learning block which is often experienced by students when approaching a new subject for the first time.

Chapters 4 and 5 survey the range of converter applications. The principles underlying signal sampling and data acquisition system configurations are described. Other functional elements, such as multiplexers and sample/holds, with which a converter must work, are introduced. No attempt is made to provide a comprehensive mathematical treatment of signal processing principles; instead, the book concentrates on those principles which are likely to have the most practical significance for designers and users of data acquisition systems. The considerations involved in interfacing converters to microprocessors are discussed.

The final chapter of the book suggests procedures for implementing practical converter applications. It provides a summary of data converter terminology

Preface

and explains how to interpret data sheet specifications and assess errors. Important points to look for, when wiring up or assembling data converter systems, are discussed.

The book gives a general outline of the range of converter devices currently available but does not attempt to list or give a detail-by-detail comparison of specific devices. Data converter devices are in a state of explosive development —the reader must look to device manufacturers' literature as the most reliable source of up-to-date information on the range of devices which are available. The book is intended to provide the reader with the basic understanding he will need if he is to readily interpret manufacturers' data sheets and application notes.

The book is directed towards practising electronics engineers who formerly regarded themselves as either purely analog or purely digital engineers, but now find that they have to get into the interface area between the two. It is also intended as a practical guide to data converters for use by students of electronic engineering and students in the measurement sciences. The practical approach which the book adopts will also make it a useful source of information for the many enthusiastic electronics hobbyists.

Self-assessment exercises are given at the end of each chapter. Full answers to these exercises are provided. The exercises are designed to provide the reader with a method of self-assessment and at the same time to reinforce and deepen his understanding of the subject.

I wish to express my gratitude to Mrs J. Daves for typing the manuscript, D. McLuskey for help in the preparation of sketches and diagrams, and J. Anderson for help in connection with the experimental work and the wiring up of some of the systems described in the text.

G. B. Clayton

1 Analog to Digital and Digital to Analog Conversion

1.1 USEFULNESS OF CONVERTERS

In many of the real world systems studied by scientists and engineers the system parameters are continuously changing quantities (analog variables) and when electronic measurement techniques are used data is derived in analog form as the electrical output signal of a transducer. It is perfectly possible to process, manipulate and store analog data using a purely analog electronic system: negative feedback techniques can make analog systems perform very precisely.[1] However, the accuracy of a purely analog system is often not usable because of the difficulty of reading, recording and interpreting analog data with high accuracy. Also, when large amounts of analog data are involved the task of analysis and storage assumes mammoth proportions.

Digital electronic systems can be made to process rapidly and accurately, to manipulate and to store large amounts of data. The advent of low cost digital microprocessor systems has drastically reduced the cost of implementing digital data processing. Microprocessors make it possible to extend the use of electronic digital techniques into areas where they were formerly not considered practicable because of economic considerations. However, digital circuits only operate on digital data—the scientist who wishes to avail himself of the power of digital techniques must first transform his analog data into digital form. A system called an *analog to digital converter*, ADC, is used to perform this function and a system called a *digital to analog converter*, DAC, performs the opposite type of conversion, transforming digital data into analog form. ADCs and DACs provide the essential interface which is required between analog and digital systems. The ADC allows a digital system to take in information from an outside analog system—the digital system can then rapidly process and analyse this information. The DAC allows the results of such analyses to be communicated back to the analog system perhaps to modify or control its action.

ADC and DAC systems are not new but until comparatively recently they were expensive to implement and have consequently been regarded as rather specialised pieces of equipment. The advent of low cost monolithic IC converter devices has changed this position: they make a wide range of versatile signal processing techniques economically available to the measurement scientist who is prepared to invest some of his time in familiarising himself with the latest devices and their capabilities. The material which follows is intended to serve as a first step in such a familiarisation exercise.

Rather than attempt a survey of available converter devices, which, in a fast-changing area such as this, would be doomed to obsolescence even before it was completed, we concentrate on the principles underlying the most popular conversion techniques. Specific IC devices are mentioned, but only as a means of relating the discussion to practical circuits which you can experimentally evaluate for yourself—there is no substitute for 'hands on' practical experience in this type of learning exercise. However, it should not be assumed that the devices referred to are the only ones, nor indeed the 'best' ones available (from a cost/performance standpoint). The choice of the 'best' converter for a specific application can only be made from a thorough study of the manufacturers' latest product guides—this book will prepare you for such a study.

1.2 DIGITAL CODES

An understanding of the operation of any DAC or ADC requires an acquaintance with the techniques which are used to represent digital numbers. In electronic systems digital numbers are represented by the presence or absence of fixed voltage levels, the numbers used are all basically binary. Each unit of information or *bit* as it is called has one of two possible states, OFF, false or 0 and ON, true or 1. The two states are represented by two voltage levels; the levels occur at the outputs of digital circuits or are applied to their inputs. Groups of levels are called *words*. they may appear simultaneously in parallel on groups of inputs or outputs, or in serial (in time sequence) at a single input or output.

The distinction between serial and parallel digital data transmission is shown diagrammatically in figure 1.1 In serial transmission the bit which occurs first in the time sequence is called the *most significant bit*, MSB, and the bit which occurs last is called the *least significant bit*, LSB. All bits in the serial word occur on the line for equal periods of time. In parallel data transmission the order of significance of the bits carried by the different lines must be specified. Serial data transmission has the advantage of economy and simplicity (it only requires a single line) but it is slow. Parallel data transmission has the advantage of greater speed since all bits are transmitted and processed simultaneously. In serial transmission information is transmitted a bit at a time; the time taken to transmit a word depends upon the length of the word (the number of bits in it).

The number of different words that can be formed from a sequence of bits is clearly dependent on how many bits there are in the sequence. Thus, 4 different words can be formed from 2 bits, 8 from 3 bits, 16 from 4 bits, 2^n words from n bits. If a digital word is to represent a number, a code must be used which defines a one-to-one correspondence between each word and a number. It is of

Analog to Digital and Digital to Analog Conversion

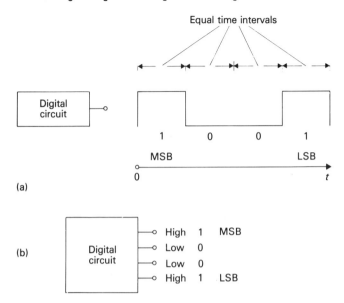

Figure 1.1 Serial and parallel digital data transmission of word 1001: (a) a serial 4-bit digital word; (b) a parallel 4-bit digital word

course possible to define quite arbitrary codes but the code which is most commonly used is *natural binary*. In the natural binary code each bit in a digital word carries a weight or multiplier which is a power of 2, the power being determined by the position of the bit in the word. Consider the digital word 1001; if this word is used to represent a number with the natural binary code we have

	Bit Weights		
MSB			LSB
2^3	2^2	2^1	2^0
1	0	0	1

and

$$1 \times 2^0 + 0 \times 2^1 + 0 \times 2^2 + 1 \times 2^3 = 9$$

Thus the digital word 1001 represents the natural binary number 1001 which is equivalent to the decimal number 9.

In converter applications it is convenient to use a fractional code in which each digital word is used to represent a fraction of full scale, where full scale refers to the maximum value of the analog input voltage in the case of an ADC or the maximum value of the analog output in the case of a DAC. In an n-bit natural binary fractional code the MSB, bit 1, is made to carry a weight of 2^{-1} ($\frac{1}{2}$) and the LSB (bit n) a weight of 2^{-n}. The decimal fraction corresponding to a particular natural binary fractional word is found by simply adding up the weights of the non-zero bits. Table 1.1 is given as an example; it shows the 16

Table 1.1 Fractional binary codes

Decimal fraction	Binary fraction	MSB ($\times 1/2$)	Bit 2 ($\times 1/4$)	Bit 3 ($\times 1/8$)	Bit 4 ($\times 1/16$)
0	0.0000	0	0	0	0
$1/16 = 2^{-4}$ (LSB)	0.0001	0	0	0	1
$2/16 = 1/8$	0.0010	0	0	1	0
$3/16 = 1/8 + 1/16$	0.0011	0	0	1	1
$4/16 = 1/4$	0.0100	0	1	0	0
$5/16 = 1/4 + 1/16$	0.0101	0	1	0	1
$6/16 = 1/4 + 1/8$	0.0110	0	1	1	0
$7/16 = 1/4 + 1/8 + 1/16$	0.0111	0	1	1	1
$8/16 = 1/2$ (MSB)	0.1000	1	0	0	0
$9/16 = 1/2 + 1/16$	0.1001	1	0	0	1
$10/16 = 1/2 + 1/8$	0.1010	1	0	1	0
$11/16 = 1/2 + 1/8 + 1/16$	0.1011	1	0	1	1
$12/16 = 1/2 + 1/4$	0.1100	1	1	0	0
$13/16 = 1/2 + 1/4 + 1/16$	0.1101	1	1	0	1
$14/16 = 1/2 + 1/4 + 1/8$	0.1110	1	1	1	0
$15/16 = 1/2 + 1/4 + 1/8 + 1/16$	0.1111	1	1	1	1

decimal fractions that can be represented by a 4-bit binary fraction. Note that in an n-bit converter using the natural binary fractional code a digital word in which all n bits are '1' corresponds to $1 - 2^{-n}$ of full scale, that is, full scale less one LSB.

1.3 CONVERSION RELATIONSHIPS

The relationships which exist between inputs and outputs in ideal converters may be shown graphically. An example is given in figure 1.2; the use of the natural binary code is assumed and for simplicity a conversion involving only 3 bits is considered. In a 3-bit DAC the digital input word can take on 8 different states giving rise to 8 different analog output signals ranging from 0 to 7/8 full scale. No other levels can exist; the graph (figure 1.2a) is a bar graph. Note that the maximum output signal produced by a DAC is always one LSB less than nominal full scale.

The analog input signal applied to an ADC must be assumed capable of taking on all values up to full scale, but in a 3-bit converter only 8 different states of the digital output word are possible. The continuum of analog input values must be partitioned (quantised) into 8 discrete ranges, as shown in figure 1.2b. All analog input values within a particular range give rise to the same digital output word corresponding to the digital output appropriate for

Analog to Digital and Digital to Analog Conversion

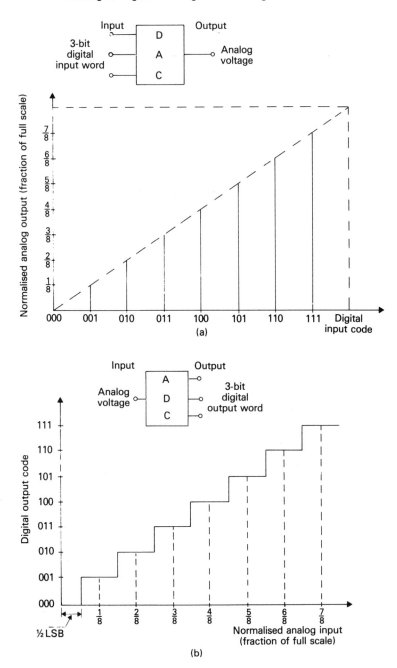

Figure 1.2 Ideal conversion relationships for 3-bit converters using natural binary code: (a) ideal 3-bit DAC; (b) ideal 3-bit ADC

the mid-range analog value. Transitions in the digital output are assumed to take place at analog input values which are $\pm \frac{1}{2}$ the LSB value either side of the mid-range value. Because of the necessity for quantisation, there is in any analogue to digital converter an inherent uncertainty (quantisation uncertainty) in the digital output of $\pm \frac{1}{2}$ LSB.

You should attempt to visualise the form of the ideal converter relationships as the number of bits is increased. The greater the number of bits the smaller is the weighting of the least significant bit and there is theoretically a corresponding greater resolution and accuracy in the conversion process. There are practical difficulties associated with any mechanisation of a conversion which means that real converters do not accurately attain their theoretical resolution limits. It is convenient to defer a treatment of practical converter errors to a later section (see chapter 6).

It should be borne in mind that natural binary is but one of many possible digital codes and in some converter applications the use of an alternative code is more convenient. In the natural binary code several bits are required to change simultaneously at various stages of the counting sequence. For example, in the 4-bit natural binary code of table 1.1, it can be seen that all 4 bits must change simultaneously in the transition between the decimal fraction 7/16 and 8/16. In a practical converter using this code it is clear that if the MSB involved in this transition were to change slightly before the other less significant bits then the code would be grossly in error for a brief period. Some converters use the so-called *Gray code* to eliminate the possibility of grossly inaccurate intermediate codes.

The Gray code is a binary code in which transitions between adjacent codes involve the change of only 1 bit at a time. Bit position does *not* signify a weighting in Gray code but each digital word in the code is still taken to represent a unique part of the analog range. 4-bit natural binary and Gray codes are compared in table 1.2.

1.4 DECIMAL CODES

Digital circuits use a form of binary arithmetic; they can only process data if it is coded in binary form. However, most people are only familiar with decimal arithmetic. To overcome this problem a digital code in which decimal numbers are encoded in binary form is often used in systems or instruments with which people interact directly. Digital voltmeters normally provide a decimal display. They use ADCs which use a decimal code. The user of a digital calculator feeds decimal numbers into the calculator, the calculator codes the numbers into binary form, performs calculations using binary arithmetic and decodes the results of the calculations back into decimal form for display.

The decimal number system uses the 10 digits 0, 1, 2, . . ., 9; if a decimal number is to be encoded in binary form each digit must be represented by the unique state of a digital word. The smallest number of binary bits which can be

Analog to Digital and Digital to Analog Conversion

Table 1.2 Comparison of 4-bit binary and Gray codes

Decimal fraction	Gray code	Binary Code
0	0 0 0 0	0 0 0 0
1/16	0 0 0 1	0 0 0 1
2/16	0 0 1 1	0 0 1 0
3/16	0 0 1 0	0 0 1 1
4/16	0 1 1 0	0 1 0 0
5/16	0 1 1 1	0 1 0 1
6/16	0 1 0 1	0 1 1 0
7/16	0 1 0 0	0 1 1 1
8/16	1 1 0 0	1 0 0 0
9/16	1 1 0 1	1 0 0 1
10/16	1 1 1 1	1 0 1 0
11/16	1 1 1 0	1 0 1 1
12/16	1 0 1 0	1 1 0 0
13/16	1 0 1 1	1 1 0 1
14/16	1 0 0 1	1 1 1 0
15/16	1 0 0 0	1 1 1 1

used to form the 10 different digital words is 4. A 4-bit word has 16 possible states any 10 of which can in principle be used to represent the 10 digits of the decimal system. The number of possible codes is

$$16 \times 15 \times 14 \times 13 \times 12 \times 11 \times 10 \times 9 \times 8 \times 7 = \frac{16!}{6!} = 2.9 \times 10^{10}$$

about 30 billion possibilities! There is no need to remember this many—only a limited number of decimal codes are in use. Indeed, the question may well be asked 'Why not standardise on one decimal code?'. The reasons associated with the development of a variety of codes are to do with the relative ease and efficiency with which available digital devices can perform manipulations on data coded in different ways.

Examples of two decimal codes which are used in converters are given in table 1.3. The 8.4.2.1 code is the decimal code which is most frequently used in A/D converters; it is commonly referred to simply as *binary coded decimal*,

Data Converters

Table 1.3 the 8.4.2.1 and 2.4.2.1 decimal codes

Decimal number	8.4.2.1 BCD $(\times 8)(\times 4)(\times 2)(\times 1)$				2.4.2.1 BCD $(\times 2)(\times 4)(\times 2)(\times 1)$			
0	0	0	0	0	0	0	0	0
1	0	0	0	1	0	0	0	1
2	0	0	1	0	0	0	1	0
3	0	0	1	1	0	0	1	1
4	0	1	0	0	0	1	0	0
5	0	1	0	1	0	1	0	1
6	0	1	1	0	0	1	1	0
7	0	1	1	1	0	1	1	1
8	1	0	0	0	1	1	1	0
9	1	0	0	1	1	1	1	1

BCD, because the weights of the bit positions are the same as in the natural binary number system. Note that the 6 upper states of the 4-bit quad are not used and are simply discarded. In the 2.4.2.1 code, which is used in some converters, the bit in the MSB position has a weight of 2 instead of the usual 8; the code has the advantages of having all '1's for (full scale − LSB) and D/A converters using the code require a smaller range of resistance values (see chapter 2).

A/D converters which are designed to give a decimal read-out require a binary quad for each output digit. Table 1.4 is given as an example; it shows BCD coding of some of the decimal fractions in the range 0 to 0.99. Note that in an A/D converter which provides 2 decimal digits at its output the magnitude of the analog input corresponding to the LSB of information presented at the output is 0.01 of full scale. The magnitude of the LSB in an 8-bit binary coded converter is $1/(16 \times 16) = 0.0039$ full scale. In a sense BCD is wasteful since it only uses 10 of the possible 16 states of a quad; a BCD quad provides only 10/16 of the resolution of a natural binary quad.

1.5 BIPOLAR CODES

The converter codes discussed thus far have related the normalised magnitudes of an analog variable, (fraction of full scale) to digital words, but no regard has been taken of the polarity of the analog signal. A variety of codes have been devised for use in conversions involving analog signals of both polarities. In these bipolar codes a digital word is made to carry information about both the magnitude and polarity of the analog signal.

Examples of binary codes which are often used in bipolar conversions are shown in table 1.5. At first sight of these codes you may be tempted to decide that the seeming jumble of '0's and '1's is far too complicated for you to decipher. However, if you study the codes a little more closely you will see

Analog to Digital and Digital to Analog Conversion

Table 1.4 Examples of 2-digit BCD weighting

Decimal fraction	BCD code	
	MSQ (× 1/10) × 8 × 4 × 2 × 1	2nd Quad (× 1/100) × 8 × 4 × 2 × 1
0.00 = 0.00 + 0.00	0 0 0 0	0 0 0 0
0.01 = 0.00 + 0.01	0 0 0 0	0 0 0 1
0.02 = 0.00 + 0.02	0 0 0 0	0 0 1 0
0.03 = 0.00 + 0.03	0 0 0 0	0 0 1 1
.		
.		
.		
0.09 = 0.00 + 0.09	0 0 0 0	1 0 0 1
0.10 = 0.10 + 0.00	0 0 0 1	0 0 0 0
0.11 = 0.10 + 0.01	0 0 0 1	0 0 0 1
.		
.		
.		
0.30 = 0.30 + 0.00	0 0 1 1	0 0 0 0
.		
.		
.		
0.90 = 0.90 + 0.00	1 0 0 1	0 0 0 0
0.91 = 0.90 + 0.01	1 0 0 1	0 0 0 1
.		
.		
.		
0.98 = 0.90 + 0.08	1 0 0 1	1 0 0 0
0.99 = 0.90 + 0.09	1 0 0 1	1 0 0 1

that the principles which underly their formulation are not all that complicated.

In a *bipolar* code 1 bit, that in the 'MSB position' in the codes shown, carries the sign information. In the *sign + magnitude* code the bits after the sign bit give the magnitude coded in natural binary; sign magnitude BCD is the code used extensively in bipolar DVMs.

The *offset binary* code is simply a natural binary counting sequence which is interpreted in terms of an offset of the analog scale which it is used to represent. The zero of the natural binary counting sequence (0000) is used to represent analog full scale negative, the count 1000 represents analog zero and the full count 1111 represents + 7/8 analog full scale (full scale − LSB). Offset binary, as will be shown when converter circuits are discussed, is one of the easiest bipolar codes to implement.

Data Converters

Table 1.5 Commonly used bipolar codes

Number	Decimal fraction	Sign + magnitude	Two's complement	Offset binary
+7	+7/8	0111	0111	1111
+6	+6/8	0110	0110	1110
+5	+5/8	0101	0101	1101
+4	+4/8	0100	0100	1100
+3	+3/8	0011	0011	1011
+2	+2/8	0010	0010	1010
+1	+1/8	0001	0001	1001
0	0+	0000	0000	1000
0	0−	1000	(0000)	(1000)
−1	−1/8	1001	1111	0111
−2	−2/8	1010	1110	0110
−3	−3/8	1011	1101	0101
−4	−4/8	1100	1100	0100
−5	−5/8	1101	1011	0011
−6	−6/8	1110	1010	0010
−7	−7/8	1111	1001	0001
−8	−8/8		(1000)	(0000)

The *two's complement* code, as can be seen, is a natural binary code for values of the analog variable from 0 to full scale positive. The code for the analog values 1/8 full scale to full scale negative is obtained by complementing in turn the codes for the values 0 to +7/8 full scale. Complementing means simply changing the state of all bits.

Table 1.5 gives bipolar codes involving 4 bits but each code is readily extended to any number of bits by using the principles on which the codes are founded which are outlined above. It should be noted that in any n-bit bipolar code 1 bit is always used for sign information, the remaining $n - 1$ bits carry the magnitude information. The analog range (± full scale) of an n-bit bipolar code is double the range covered by an n-bit unipolar code but the ratio

$$\frac{\text{analog magnitude of LSB}}{\text{full scale analog magnitude}}$$

is $2^{-(n-1)}$ for the bipolar code and not 2^{-n} as it is for a unipolar code. The resolution of an n-bit bipolar code is twice as coarse as that of an n-bit unipolar code. that of an n-bit unipolar code.

Bipolar codes do require rather more care in interpretation than unipolar codes. In table 1.5 it has been assumed that an increase in the digital number corresponds to a positive increase in the analog variable, but this is not always so—it depends on the details of the converter circuitry. In some converters an

Analog to Digital and Digital to Analog Conversion

increase in the digital numbers corresponds to an increasingly negative value of the analog variable and if this is the case the bipolar codes should be reinterpreted by changing the signs of all the decimal fractions.

The treatment of codes has by no means covered all codes in use but it has, hopefully demonstrated the essential role of a code in any A/D or D/A conversion.

1.6 SELF-ASSESSMENT EXERCISES

1. Fill in the blanks in the following statements.
In the terminology of digital electronics

   (a) A bit is a ....
   (b) A bit has .... states represented by ....
   (c) A digital word is ....
   (d) In serial transmission of digital data bits are transmitted ....
   .... , in parallel transmission bits are transmitted ....
   (e) 5 bits can be used to form .... different words.

2. Which of the following statements are true?
   (a) A 6-bit digital word requires 6 data lines for serial transmission.
   (b) Parallel data transmission is faster than serial transmission.
   (c) Parallel data transmission is simpler than serial transmission.
   (d) The Gray code is a positional weighted code.
   (e) The BCD code is a positional weighted code.
   (f) There is a possible error in the digital output of an A/D converter even if the converter is ideal.

3. Give the decimal equivalent of
   (a) the natural binary numbers

   1001; 10111; 110

   (b) the natural binary fractions

   .11; .001; .010; .000001; .01010

4.

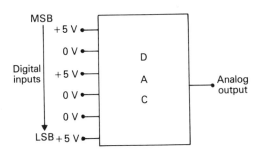

12   Data Converters

The DAC shown in the sketch has a nominal full scale analog output of 10 V. The digital signal +5 V represents binary '1' and 0 V represents binary '0'.

(i) What is the value of the analog output signal assuming the DAC uses: (a) a natural binary code, (b) an offset binary code, (c) a two's complement code.

(ii) What digital inputs cause the magnitude of the DAC's analog output to have: (a) its largest value, (b) its smallest non-zero value? What are these largest and smallest analog output voltages?

5.

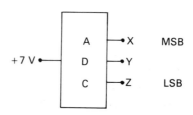

The sketch shows a 3-bit natural binary coded A/D converter. The nominal full scale analog input signal is 10 V. Assuming +5 V represents binary '1' and 0 V represents binary '0', what voltage levels do you expect to observe at points X, Y, Z?

6. (i) Show graphically the input output relationship which exist for ideal 4-bit D/A and A/D conversion (similar to figure 1.2). Assume: (a) a natural binary code, (b) an offset binary code, (c) a BCD code. What is the magnitude of the values of the analog input signals which give rise to the digital output code 0111 in the A/D conversions.

7. Give the BCD converter codes for the decimal fractions: (a) 0.17, (b) 0.56, (c) 0.95.

8. Give the decimal fractions which are represented in the BCD converter code by the following digital words

(a) 0011
(b) 0101
(c) 1101
(d) 01011000
(e) 11001011.

Analog to Digital and Digital to Analog Conversion

9. Give the decimal fractions corresponding to the following digital words

   011; 11010; 01110; 100001

   (a) in the offset binary converter code,
   (b) in the sign + magnitude converter code,
   (c) in the two's complement converter code.

2 Digital to Analog Converters

In the discussion of conversion principles presented thus far the function of a DAC has been established as that of providing an analog output signal in response to a digitally coded input signal. The basic circuit principles underlying the implementation of this function are not difficult to understand and can readily be demonstrated in a simple but convincing manner.

2.1 IMPLEMENTING A DAC FUNCTION

Connect up, or simply consider, the circuit arrangement given in figure 2.1. It consists of a reference voltage source and a set of binary weighted resistors,

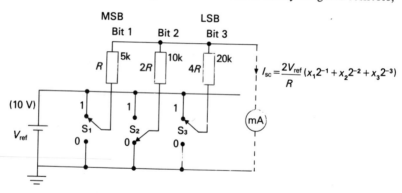

Figure 2.1 D/A conversion with binary weighted resistors

each resistor having an associated switch. Switch positions are taken as representing values of binary inputs. If a switch is in the state designated '1' V_{ref} causes a current to flow through the resistor associated with that switch. The sum of all switched current contributions is the short circuit output current of the network; it can be measured by a low resistance milliammeter to give an analog reading corresponding to the binary coded digital input.

Digital to Analog Converters

15

The MSB (bit 1) switch makes a contribution V_{ref}/R to the short circuit output current, bit 2 makes a contribution $V_{ref}/2R$ and bit 3, which in figure 2.1 is the LSB, makes a contribution $V_{ref}/4R$. Using $V_{ref} = 10$ v and $R = 5$ kΩ as suggested makes the LSB contribution $10/20 = 0.5$ mA and with all bits on (binary input 111), the short circuit output current is 3.5 mA (7/8 full scale where full scale is 4 mA).

You should measure, or simply calculate, the value of the short circuit output current for all possible switch positions corresponding to the 8 different states of the 3-digit binary input word. Graph the relationship between output current and digital input word and compare with figure 1.2a.

A digital to analog conversion involving a digital input word with more than 3 bits can be implemented using the principles outlined above by simply adding an extra switch and resistor for each extra bit. Thus an n-bit natural binary DAC would require n binary weighted resistors values $R, 2R, 4R, \ldots, 2^{n-1}R$. An expression for the short circuit output current developed by such a network would be

$$I_o \text{(short circuit)} = \frac{2V_{ref}}{R}\left(x_1 2^{-1} + x_2 2^{-2} + x_3 2^{-3} + \ldots + x_n 2^{-n}\right) \quad (2.1)$$

where $x_i = 1$ if S_i is switched to the high state or $x_i = 0$ if S_i is switched to the low state.

2.2 ANALOG READ-OUT

There are a variety of techniques possible for reading out the analog output signal produced by a binary weighted resistor network. An operational amplifier can be used to give a current sum to voltage conversion as shown in figure 2.2a. Here the analog output signal is in the form of a low output impedance voltage signal which can be scaled by choice of R_f and has a value determined by the relationship

$$V_o = \frac{-2V_{ref}}{R} R_f \left(x_1 2^{-1} + x_2 2^{-2} + x_3 2^{-3} + \ldots + x_n 2^{-n}\right) \quad (2.2)$$

Note that the analog output polarity in this case is negative and goes more negative as the value of the digital input word is increased.

Alternatively an operational amplifier can be used in the high input impedance follower configuration as shown in figure 2.2b. This arrangement allows the open circuit output voltage produced by the resistor network to be read out at low impedance as

$$V_o = \frac{2^n}{2^n - 1} V_{ref} \left(x_1 2^{-1} + x_2 2^{-2} + x_3 2^{-3} + \ldots + x_n 2^{-n}\right) \quad (2.3)$$

16 Data Converters

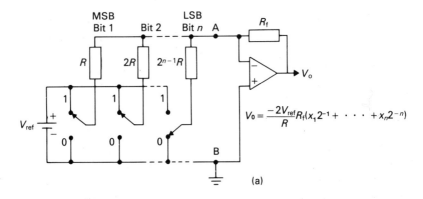

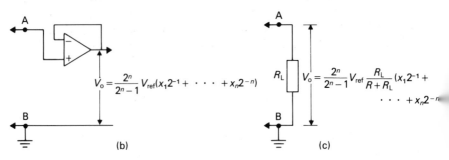

Figure 2.2 Read-out of analog output signal: (a) operational amplifier converts short circuit output current to low impedance voltage; (b) open circuit output voltage read-out; (c) a loaded network

Loading the binary network with a load resistor R_L as shown in figure 2.2c, gives rise to an output voltage developed across the load of value

$$V_o = \frac{2^n}{2^n - 1} V_{ref} \frac{R_L}{R_o + R_L} \left(x_1 2^{-1} + x_2 2^{-2} + x_3 2^{-3} + \ldots + x_n 2^{-n} \right) \quad (2.4)$$

where R_o is the effective output resistance of the binary network, $R \parallel 2R \parallel 4R \parallel \ldots \parallel 2^{n-1}R$.

$$R_o = \frac{2^{n-1}}{2^n - 1} R \approx \frac{R}{2} \quad \text{for } n > 4$$

Note that when the resistor network is loaded any change in load inevitably influences the analog output signal.

2.3 USE OF R-2R LADDER NETWORKS

In concept, weighted resistor networks provide the simplest and most direct method of performing a D/A conversion. However, when many bits of digital

Digital to Analog Converters

information are involved, the weighted resistor network has the disadvantages of requiring a large range of resistor values. A 10-bit converter would require resistor values in the range $2^9:1$ (512:1) and the MSB resistor would need to be of very close tolerance if it were not to introduce errors as big as the LSB value. In a 10-bit converter the size of the LSB is only $1/2^9 \times 100 \approx 0.2$ per cent of the MSB. The MSB resistor value would need to be accurate to better than ± 0.2 per cent if it were not to introduce an error as big as the LSB.

The difficulties associated with a requirement for a wide range of precision binary weighted resistors are overcome in many practical converters by the use of a resistor ladder network of the form shown in figure 2.3. The network

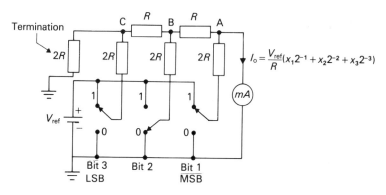

Figure 2.3 R–2R ladder network gives binary bit weighting

maintains a binary weighting of bit currents but has only 2 resistor values: it is called an R–$2R$ ladder network.

In figure 2.3 a 3 stage R–$2R$ network is considered for the sake of simplicity but the principles involved in the action of the network are readily extended to any number of stages. First notice that, regardless of the number of stages, the effective output resistance of the network (looking back to the left in figure 2.3) is R. The output resistance looking back into the circuit at point C is $2R \parallel 2R = R$, and at point B is R plus R in series in parallel with $2R$, namely $2R \parallel 2R = R$ and so on regardless of the number of stages.

The output voltage produced by the network can be derived using the principle of superposition as the sum of the effects of the individual bits acting separately. The effect of each bit at the output is most readily found by deriving its Thévenin equivalent, the process is shown in figure 2.4. In deriving the Thévenin equivalent for a particular bit all bit switches except that for the bit under consideration are immagined in the '0' state. It can be seen that the MSB (bitone) makes a contribution $V_{ref}/2$ to the open circuit output voltage, bit 2 makes a contribution $V_{ref}/4$ and bit 3 a contribution $V_{ref}/8$.

In the more general case of an R–$2R$ network with n stages used for an n-bit D/A conversion an expression for the open circuit output voltage is

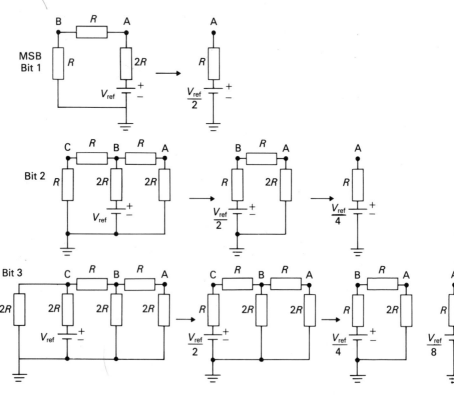

Figure 2.4 Thévenin equivalent circuits for each separate bit in figure 2.3

$$\underset{\text{(open circuit)}}{V_o} = V_{\text{ref}}\left(x_1 2^{-1} + x_2 2^{-2} + \ldots + x_n 2^{-n}\right) \tag{2.5}$$

where as before $x_i = 1$ if switch S_i is in the high state or $x_i = 0$ if switch S_i is in the low state.

An expression for the short circuit output current is

$$\underset{\text{(short circuit)}}{I_o} = \frac{V_{\text{ref}}}{R}\left(x_1 2^{-1} + x_2 2^{-2} + \ldots + x_n 2^{-n}\right) \tag{2.6}$$

An expression for the output voltage of the network when loaded by a resistor R_L is

$$V_o = V_{\text{ref}} \frac{R_L}{R + R_L}\left(x_1 2^{-1} + x_2 2^{-2} + \ldots + x_n 2^{-n}\right) \tag{2.7}$$

If the analog output voltage must be available at a low output impedance an operational amplifier can be used as shown previously in figures 2.2a and 2.2b.

R-$2R$ ladder networks, because of their symmetry can be used in a variety of circuit configurations. In the arrangement shown in figure 2.5 the reference input and output lines of figure 2.3 are interchanged. A change of switch state in figure 2.5 causes very little change in the voltage level at the switch. The

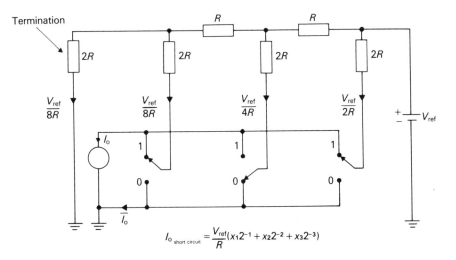

Figure 2.5 R-2R network in current switching configuration

short circuit output current produced by the simple 3-bit arrangement is determined by the relationship

$$I_o \text{ (short circuit)} = I_{ref}\left(x_1 2^{-1} + x_2 2^{-2} + x_3 2^{-3}\right) \quad (2.8)$$

The R-$2R$ network divides the input current $I_{ref} = V_{ref}/R$ into binarily related bit current components which the switches steer to either the output line or ground. Notice that a current increment equal in value to the LSB current flows through the terminating $2R$ resistor to ground. The number of bits can be increased by extending the network by simply adding extra sections to the R-$2R$ ladder. Analog Devices AD7500 series of multiplying DACs (see section 4.2) use an R-$2R$ network of the type shown in figure 2.5.

Our treatment of resistor weighting networks has by no means covered all the techniques which are used in practical converters. The R-$2R$ ladder is probably the most frequently used network for binary bit weighting but an alternative approach which is adopted in some converters is to use binary weighted resistor quads ($2R$, $4R$, $8R$, $16R$) with appropriate attenuation between the quads. The quad approach allows the proper relative quad weighting for BCD conversion to be obtained by adjustment of this interquad attenuation, a circuit configuration illustrating the use of binarily related resistor quads is given in figure 2.6.

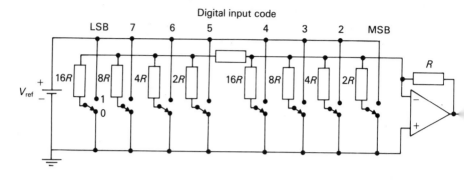

Figure 2.6 8-bit D/A converter using two equal resistance quads with attenuation to the less significant quad

The subject of resistor weighting will not be pursued further since from the DAC user's point of view a general knowledge of the basic ideas underlying the subject is all that is required. Commercially available DACs contain resistor weighting networks, but the devices can be used effectively without a detailed knowledge of the design of these networks. Practical DACs do not of course use mechanical switches, they use electronic switches which are activated in response to the high or low voltage levels which are applied to their logic inputs. Current switching techniques based on the circuit configuration of figure 2.5 (because they involve very little change in switch voltage), provide faster operation than the voltage switching of figure 2.2. Bipolar transistor current switches are used in many converters but the detailed circuitry involved in such switching arrangements need not concern the DAC user.

2.4 PRACTICAL DACS

A wide variety of DACs are available in both integrated circuit and modular form, ranging from modest 6-bit converters to very accurate 16-bit converters. Available devices differ in speed, accuracy and the range of performance options which they provide (types of digital code, analog polarity, etc.). Some devices include their own built in reference voltage while in others the reference voltage must be externally connected by the user. Devices in which the external reference voltage can be varied are referred to as *multiplying DACs*, since in these devices the analog output signal is proportional to the product of the variable reference voltage and the input digital number. Some devices produce an output current which, if required, can be converted to a low output impedance voltage by means of an externally conected operational amplifier, while others include an internal operational amplifier which is used to perform this function. The output operational amplifier in a converter, if it is used, invariably slows down the response of the converter.

No attempt will be made to survey the range of existing DAC devices: new

Digital to Analog Converters

devices continue to emerge and any list would be doomed to rapid obsolescence. The only way to satisfactorily choose a suitable DAC for a specific application is to consult the latest manufacturer's product guide and data sheets. However, the newcomer to DACs will find it a great help if before studying manufacturer's literature he first acquires a practical understanding of the operation of a specific device. Familiarity with a specific converter, gained as a result of experimentation, will make him more readily appreciate the significance of the information and performance parameters which he is likely to find included in DAC data sheets.

2.5 EXAMINING THE PERFORMANCE OF A DAC

An experimental learning exercise on DACs is best performed with a device which allows a range of different operating conditions and thereby permits the experimenter to more fully investigate the factors influencing DAC performance. Precision Monolithics multiplying D/A converter type DAC-08 is chosen for discussion here; there are of course other inexpensive integrated circuit DACs available, for example, AD7500 series or MC14086-8 — if you decide to use one of these alternative devices you will need to study its data sheet in detail first.

The DAC-08 is an 8-bit integrated circuit multiplying DAC; it is fast, it provides a range of flexible operating conditions and is inexpensive. In figure 2.7, which is extracted from the manufacturer's data sheet, the device pin connections and simplified equivalent circuit are shown. Pins 5 through to 12 are the logic inputs, the MSB pin 5 and the LSB pin 12. The logic threshold can be adjusted by means of a voltage applied to the logic threshold control, pin 1; this feature enables the device to be interfaced with all the popular logic families. If pin 1 is grounded the device responds to TTL logic levels. The device data sheet should be consulted for interfacing techniques to other families.

An internal operational amplifier, together with an external reference voltage and resistor, is used to set the value of a reference current. The reference current is divided into binarily related bit currents by an R-$2R$ ladder network and the bit currents are supplied to current switching transistors. The simplified equivalent circuit of figure 2.7 does not show the detailed switching circuitry nor does it indicate the technique used to obtain correct scaling of the LSB current increment.

The reference amplifier connections for positive, negative and bipolar reference inputs are shown in figure 2.8. Transistor T' and the current sink bit transistor $T_1, T_2, T_3, \ldots, T_8$, share a common base line driven by the output voltage of the reference amplifier. Transistor collector and emitter currents are approximately equal and the voltage $I_{ref}R$ which appears across the emitter resistor of T' drives the R-$2R$ network (compare with figure 2.5).

Feedback round the reference amplifier is returned to its non-inverting

22 **Data Converters**

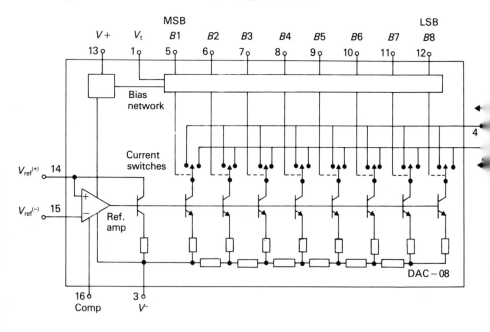

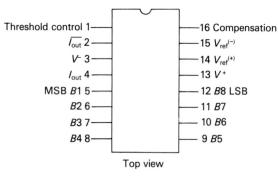

Figure 2.7 Precision Monolithics DAC-08 pin connections and simplified functional schematic

input terminal; the connection gives negative feedback because of the signal phase inversion between the base and collector of transistor T'. Assuming the reference amplifier behaves like an ideal operational amplifier, all current arriving at pin 14 is made to flow as the collector current of T' and the voltage levels at pins 14 and 15 are forced to equality. The negative reference connection of figure 2.8b in effect applies series negative feedback to the reference amplifier and is thus characterised by a high input impedance. Connections

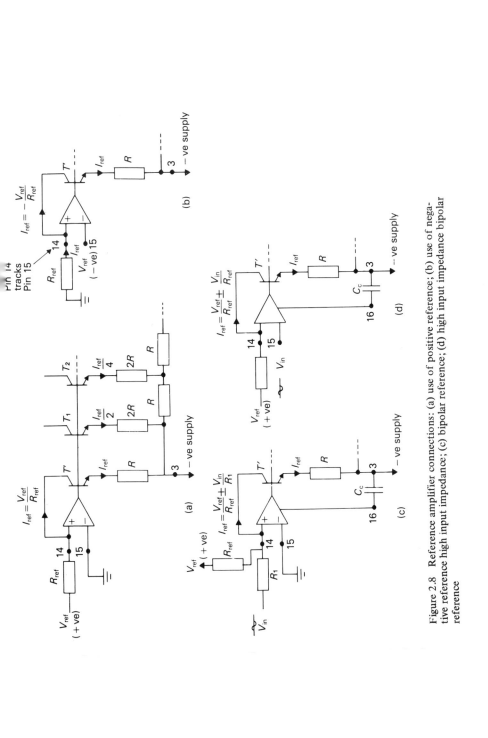

Figure 2.8 Reference amplifier connections: (a) use of positive reference; (b) use of negative reference high input impedance; (c) bipolar reference; (d) high input impedance bipolar reference

for use with variable bipolar reference inputs are obtained by d.c. offsetting the current into pin 14 and are shown in figures 2.8c and 2.8d; values used must ensure that the current direction is always into pin 14. In multiplier applications when an alternating reference signal is applied a capacitor, C_c, must be connected between pin 16 and pin 3 (the negative supply) in order to frequency compensate the reference amplifier. The value required for C_c depends on the value used for R_{ref}, the minimum recommended values of C_c are 15 pF, 37 pF and 75 pF for R_{ref} values 1 kΩ, 2 kΩ and 5 kΩ respectively.

A feature of the DAC-08 not commonly found in other devices, is that it provides two output currents, the current I_o at pin 4 and the current $\bar{I}_o$ at pin 2, currents flow into the output terminals. Bit currents, instead of being switched between a single output line and ground are switched between the I_o and $\bar{I}_o$ lines. A bit current is switched to the I_o line when its input logic terminal is in the state '1' and to the $\bar{I}_o$ line when the logic terminal has the state '0'.

Output currents have values which are determined by the relationships

$$I_o = I_{ref}\left(x_1 2^{-1} + x_2 2^{-2} + \ldots + x_8 2^{-8}\right) \tag{2.9}$$

and

$$\bar{I}_o = I_{ref}\left(\bar{x}_1 2^{-1} + \bar{x}_2 2^{-2} + \ldots + \bar{x}_8 2^{-8}\right) \tag{2.10}$$

Note that $I_o + \bar{I}_o = I_{fs}$ where I_{fs} is the full scale output current determined by the relationship

$$I_{fs} = \frac{255}{256} I_{ref}$$

Both output currents can be used simultaneously, but if one output is not required it must be connected to ground or to a current point capable of supplying the current I_{fs}. Both outputs can be converted into voltage signals by simply using an external load resistor, or if a low output impedance voltage signal is required an operational amplifier can be used as a current to voltage converter. The outputs have a wide voltage compliance: *output voltage compliance* is the range of output voltage which can be applied to an output terminal without changing the value of the output current.

As an experimental familiarisation exercise it is suggested that you connect the DAC logic inputs to the parallel outputs of an 8-bit binary counter; a suitable arrangement is shown in figure 2.9. Wire up the counter first, connect a pulse generator to the clock input and use an oscilloscope to verify that it is operating correctly. The counter LSB frequency should be half the clock frequency and each succeeding bit should have a frequency half that of the preceding bit. Next connect up the DAC and carefully check all wiring before switching on the power supplies. Note that in figure 2.9, the 12 V positive supply line is used as the DAC reference voltage supply; in a practical application a separate reference voltage would normally be used for greater accuracy.

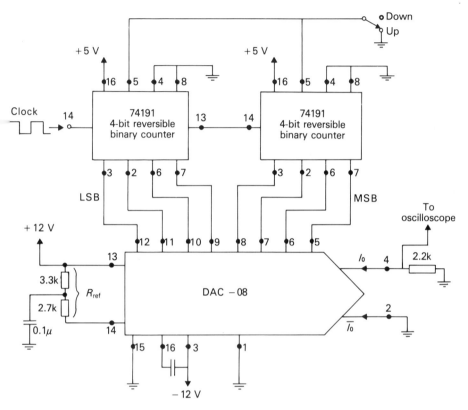

Figure 2.9 Incrementing a DAC with a binary counter

Set the clock frequency to a convenient value (say 100 kHz) and observe the analog output signal (at pin 4) with an oscilloscope. The traces given in figure 2.10 show you what you should expect to see. In the lower trace the oscilloscope vertical and horizontal controls have been adjusted so as to 'stretch out' the DAC output waveform and make the transitions between output states clearly visible. The output is incremented one LSB at a time; it returns to zero each time the counter recycles to a zero count (after 256 clock pulses). Measure the LSB output change and express it as a fraction of the full scale output.

There are many other aspects of the DAC performance that you can investigate—here are a few suggestions. Connect a second 2.2 kΩ resistor between the $\bar{I}_o$ output, (pin 2) and ground and simultaneously observe both outputs. Change the value of the reference current (by changing the value of R_{ref}) but do not exceed I_{ref} = 3 mA. Try the effect of setting the counters in the count

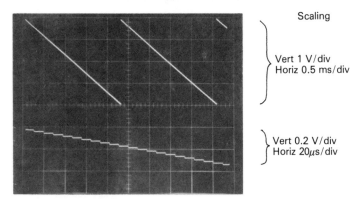

Figure 2.10 Output from DAC

down mode by applying a logical '1' instead of '0' to the counter pin 5. In each case you should observe, record and explain the effects. Another experiment that can usefully be performed is to give the DAC a bipolar output as described in the next section. Suggested components for bipolar symmetrical offset binary operation are given in figure 2.12.

2.6 BIPOLAR CODED DACS

In some applications DACs are required to produce a bipolar analog output in response to a bipolar coded digital input. Most natural binary coded DACs can readily be adapted to bipolar, offset binary coded operation. The modifications required to give a natural binary coded DACs a bipolar sign magnitude code are not so conveniently implemented. However ready built, monolithic, sign magnitude coded DACs are freely available.

2.6.1 Offset Binary Coded DACs

A bipolar output signal can be obtained from a unipolar output DAC by simply offsetting the unipolar output. The technique is illustrated in figure 2.11—it is applicable to many currently available DACs. In order to give an offset binary code an offset current $I_{ref}/2$ is simply summed with the DAC's unipolar output current I_o. The summation is performed by the operational amplifier which is normally used to convert the DAC's output current to an output voltage. The operational amplifier produces an output signal

$$V_o = \left(I_o - \frac{I_{ref}}{2}\right) R$$

where

$$I_o = I_{ref} \left(x_1 2^{-1} + x_2 2^{-2} + \ldots + x_n 2^{-n}\right)$$

Digital to Analog Converters

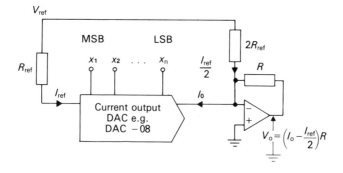

	(Sign bit) MSB	Bit values						LSB	Output as fraction of $I_{ref}R$
	x_1	x_2						x_n	$n=8$
Positive full scale	1	1	1	1	1	1	1	1	$+127/256$
Positive full scale − LSB	1	1	1	1	1	1	1	0	$+126/256$
Zero scale	1	0	0	0	0	0	0	0	0
Zero scale − LSB	0	1	1	1	1	1	1	1	$-1/256$
Negative full scale + LSB	0	0	0	0	0	0	0	1	$-127/256$
Negative full scale	0	0	0	0	0	0	0	0	$-128/256$

Figure 2.11 Bipolar output DAC with offset binary code

Substitution gives

$$V_o = I_{ref}R\,[(x_1 2^{-1} + x_2 2^{-2} + \ldots + x_n 2^{-n}) - \tfrac{1}{2}] \qquad (2.11)$$

Equation 2.11 may be used to obtain the conversion code; some code values for an 8-bit converter are tabulated in figure 2.11. Note that the LSB analog increment is $(1/256)(1/2^n)$ of the full scale analog range (full scale negative to nominal full scale positive) but this is $(1/128)(1/2^{n-1})$ of the full scale magnitude of the bipolar code.

2.6.2 DAC with a Symmetrical Offset Binary Code

DACs in which the bit ON and bit OFF output current increments are steered to separate externally acessible output pins can be configured to give a bipolar analog output without an external offset. The analog output is simply made proportional to the difference between the normal and complementary DAC output currents. In DACs like the DAC-08 in which output currents have a wide voltage compliance, a single operational amplifier can be configured to perform the current subtraction. A circuit is illustrated in figure 2.12. Pin connections and suggested component values are shown for the benefit of readers who wish to obtain an experimental familiarity with this technique.

Assuming ideal operational amplifier behaviour the output voltage is determined by the relationship

$$V_o = \left(I_o - \overline{I}_o\right) R_2$$

I_o is determined by equation 2.9 and for the 8-bit converter

$$\overline{I}_o = \frac{255}{256} I_{ref} - I_o$$

Thus

$$V_o = \left(2 I_o - \frac{255}{256} I_{ref}\right) R_2 \qquad (2.12)$$

On substituting for I_o, equation 2.12 becomes

$$V_o = I_{ref} R_2 \left[2 \left(x_1 2^{-1} + x_2 2^{-2} + \ldots x_8 2^{-8}\right) - \frac{255}{256} \right] \qquad (2.13)$$

Equation 2.13 may be used to obtain the conversion code; some values of this code are tabulated in figure 2.12. The code is symmetrical offset binary. In this code the analog output states for positive and negative outputs are symmetrically displaced about zero. Note that there is no value of the digital input for which the analog output is identically zero. Since the code is bipolar the analog magnitude of an LSB increment is twice as big as that in a unipolar code with the same number of bits. In a bipolar code half the code words are assigned to represent positive analog values, the other half represent negative analog values. The analog magnitude of an LSB increment is $1/2^n$ of the full scale analog range. In a bipolar converter the full scale analog range is the range between full scale negative and full scale positive. The LSB increment in a bipolar converter is thus $1/2^{n-1}$ of the full scale analog magnitude.

2.6.3 Sign Magnitude Coded DACs

A unipolar, current output, natural binary coded DAC can be made to produce a bipolar output voltage with sign magnitude coding. 1 bit of the DAC's digital input word (the sign bit) is made to activate a semiconductor switch which is used to control the converter's output voltage polarity. The principle of

Digital to Analog Converters

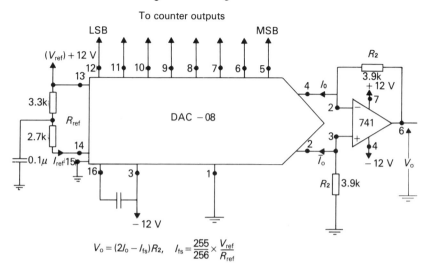

	(Sign out) MSB	Bit values						LSB	Output as fraction of $I_{ref}R_2$
	X_1	X_2	X_2	X_4	X_5	X_6	X_7	X_8	
Positive full scale	1	1	1	1	1	1	1	1	$+255/256$
Positive full scale $-$ LSB	1	1	1	1	1	1	1	0	$+253/256$
Zero scale $(+)$	1	0	0	0	0	0	0	0	$+1/256$
Zero scale $(-)$	0	1	1	1	1	1	1	1	$-1/256$
Negative full scale $+$ LSB	0	0	0	0	0	0	0	1	$-253/256$
Negative full scale	0	0	0	0	0	0	0	0	$-255/256$

Figure 2.12 Symmetrical offset binary operation of DAC-08

this switching is illustrated by the functional schematic given in figure 2.13.

Two output operational amplifiers are used in figure 2.13: amplifier A_1 acts as a current to voltage converter, A_2 acts as a current inverter (current mirror). Assuming ideal amplifier action, A_2 maintains its differential input terminals at the same potential. If a current is injected into the point X this same current must flow towards the point Y (because of the equal value resistors R'). Thus if the sign bit is in the state '0' the DAC ouput current I_o is inverted

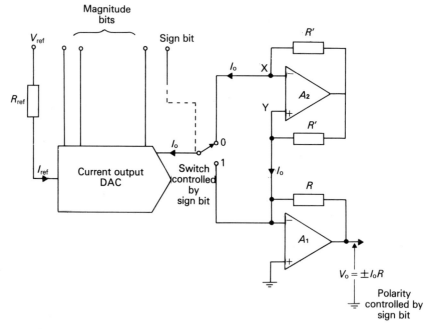

Figure 2.13 Output polarity switching gives a natural binary DAC a sign magnitude code

by A_2 before being converted to a voltage by A_1; A_1 then produces a negative output voltage. If the sign bit is '1' the DAC output current is switched directly to A_1 causing a positive output voltage to be produced. The magnitude of the output voltage is determined by the other bit states which are applied to the DAC in accordance with a natural binary code.

Output polarity switching can be used to transform any unipolar natural binary coded DAC into a bipolar system with sign magnitude code. However, it is normally more convenient to buy a ready built sign magnitude coded DAC rather than attempting externally to modify an existing natural binary coded DAC. Monolithic sign magnitude coded DACs are readily available (for example, Precision Monolithics, DAC-02, a 10-bit sign monolithic device).

2.6.4 Two's Complement Bipolar Coded DAC

The two's complement digital code is used extensively in computer and microprocessor applications. Most natural binary coded DACs can readily be modified to give a bipolar output with two's complement code. Inspection of table 1.5 reveals that a two's complement code is an offset binary code with its MSB (its 'sign bit') complemented. Thus, to implement a two's complement code with a natural binary DAC it is simply necessary to offset the DAC's output by half the full scale range and to connect an inverter to the DAC's most significant bit. The technique is illustrated in figure 2.14.

Digital to Analog Converters

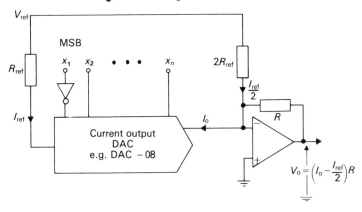

	(Sign out) MSB			Bit values				LSB	Output as fraction of $I_{ref}R$
	x_1	x_2	x_2	x_4	x_5	x_6	x_7	x_8	
Positive full scale	0	1	1	1	1	1	1	1	+127/256
Positive full scale − LSB	0	1	1	1	1	1	1	0	+126/256
Zero scale	0	0	0	0	0	0	0	0	0
Zero scale − LSB	1	1	1	1	1	1	1	1	−1/256
Negative full scale + LSB	1	0	0	0	0	0	0	1	−127/256
Negative full scale	1	0	0	0	0	0	0	0	−128/256

Figure 2.14 Bipolar output DAC with two's complement code

The inverter on the MSB digital input makes the natural binary, current output DAC produce a current which is related to bit values by the expression

$$I_o = I_{ref} \left(\bar{x}_1 2^{-1} + x_2 2^{-2} + \ldots + x_n 2^{-n} \right) \quad (2.14)$$

With the DAC's output current offset by $I_{ref}/2$ the output operational amplifier, configured as a current to voltage converter, produces an output voltage

$$V_o = \left(I_o - \frac{I_{ref}}{2} \right) R$$

Substitution for I_o from equation 2.14 gives

$$V_o = I_{ref}R\,[(\bar{x}_1 2^{-1} + x_2 2^{-2} + \ldots + x_n 2^{-n}) - \tfrac{1}{2}] \qquad (2.15)$$

Equation 2.15 defines the two's complement conversion code; some code values for an 8-bit DAC are tabulated in figure 2.14.

2.7 ADDING A DIGITAL DATA STORAGE REGISTER TO A DAC

The analog output produced by a DAC continuously reflects the state of its logic inputs. If input logic levels change there is a corresponding change in analog output. In some DAC applications (for example, data distribution) new digital data continuously appears but it is desired that the DAC output should represent the analog value of the digital data present at a particular instant in time. Thereafter the DAC is required to hold the analog value corresponding to that digital data until such time as it is instructed to acquire a new value. The functional operation described can be implemented by supplying a DAC's logic input through some form of digital data storage register or latch. This operation is illustrated in figure 2.15.

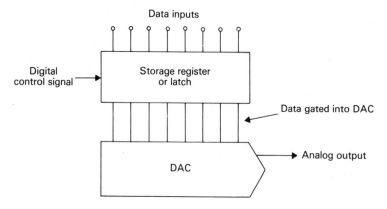

Figure 2.15 DAC with buffer register

There are a variety of both TTL and CMOS MSI devices available which are suitable for implementing the digital data storage function shown in figure 2.15 (for example, TTL 74LS363, CMOS 4099; readers are advised to consult digital device catalogues for full information). Also there are monolithic D/A converter devices available which incorporate an input digital data latch formed in on the same chip as the converter circuitry (for example, Analog Devices AD7524, AD558, Motorola MC6890). The intense competition between the manufacturers of data converter devices results in the continual emergence of new devices. The new devices are all directed towards simplifying the system

Digital to Analog Converters 33

designer's task by reducing the number of circuit packets which he needs to use to implement his data system. New devices, in general, provide more convenient, lower cost design approaches and this makes it important for the designer to keep in touch with the latest device developments. Unfortunately this task can be rather time consuming because of the proliferation of new devices.

The operation of the system outlined in figure 2.15 is as follows. The storage register is operated by a digital control signal which is sometimes called an *enable* or *strobe* signal. In one state of the control signal the register may be thought of as transparent—the data present at its inputs is simply routed to its outputs. When the control signal is switched to its other state the data present at the register outputs at the instant of switching is held permanently stored there and is unaffected by changes in the input data. The analog output of the DAC is then similarly unaffected by changes in digital input data.

In a practical system there is a limit to the rate at which the control signal can instruct the DAC to acquire new analog output values. The analog output signal produced by a DAC takes a finite time to settle to a new value in response to a change in its digital logic levels. It is this output settling time which is the main factor governing the rate at which the DAC output can be updated to new values. Digital logic circuits themselves take a finite time to change but in general the settling time associated with the analog sections of a DAC is at least an order of magnitude slower than the response time of the digital signals. Digital response times do not therefore have any significant effect on the rate at which a DAC's output can be updated. The main importance of the response times associated with the digital sections of a DAC lie in their effect in determining so called *glitches*. These are anomalies which occur in a DAC's output because of unequal turn on and turn off times of the bit currents (see section 2.8).

2.7.1 Registers Allow a DAC to Accept Serial Input Data

The basic DACs discussed in this chapter require that their digital input data be in parallel form. However, if digital data is to be transmitted over considerable distances it can be advantageous to transmit the data in serial form (fewer logic lines). If serial data is to be converted into an analog signal it must first be converted into parallel form before applying it as a DAC's digital input.

Shift registers are functional elements which are commonly used to convert the format of digital data (serial to parallel, parallel to serial). The combination of a shift register and a storage register used preceding a DAC allows the DAC to accept its digital data in serial form; a functional arrangement is illustrated in figure 2.16.

The sequence of operation involved in figure 2.16 is as follows. As each bit of a serial data word occurs on the input line it is clocked into the shift register. The data is shown fed in from the left; the action of the shift register

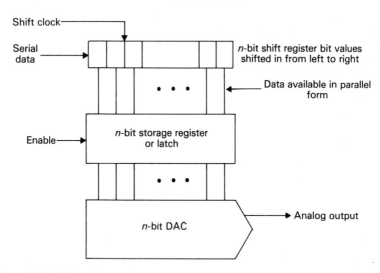

Figure 2.16 Shift register and buffer register allow DAC to accept serial data

is such that as each new bit is fed into it previously occurring bits are moved one storage position to the right. A serial n-bit word is clocked into the n-bit shift register by n clock pulses. The storage register holds the DAC's digital input constant while the shift register is being loaded with a new data word. The DAC's analog output then corresponds to a previously received data word. Once a new serial word has been loaded into the shift register it is available in parallel form and can be transferred at any time to the storage register thus updating the DAC's output.

Suitable MSI digital devices for implementing the above operation are to be found in both TTL and CMOS product lists. There are also single packet devices available which incorporate both the data format conversion and the storage function (for example, CMOS4094).

2.8 SETTLING TIME, TRANSIENT ERRORS, GLITCHES

An ideal DAC would produce a 'clean' instantaneous step change in its analog output in response to a change in its digital logic levels. Real DACs depart somewhat from this ideal mode of behaviour. The output of a practical device takes a finite time for its output to change and accurately attain a new value when its bit states are changed. The time taken for the output to change and settle within a specified error band (namely $\pm \frac{1}{2}$ LSB) about its final value is called the *settling time*. Settling time depends on the magnitude of the bit currents involved in a change, an MSB change takes longer to settle than an LSB change. Settling time is governed by the dynamic characteristics of a DAC's analog output circuitry.

Some DAC systems do not use an output operational amplifier—output

Digital to Analog Converters

current is converted to an output voltage by simply connecting an external load resistor to the DAC output current line. In systems of this type output settling time is governed by the CR time constant associated with the external load. Capacitance includes the DAC output capacitance and any stray capacitance in parallel with the external load resistor. In DAC systems which use an output operational amplifier to provide a low output impedance voltage signal it is the settling time of the operational amplifier which is a major influence on system settling time. [1]

DAC settling time is prolonged by the occurrence of sometimes large transient errors in output current which occur at the instant of an input digital code change. These transient errors are called *glitches*. Glitches are particularly undesirable when DACs are used to drive the beam deflection circuits for a CRT which is being used to provide a visual display. Glitches cause a reggedness in displayed patterns.

Glitches are due to unequal turn ON and turn OFF times of the switches which are used to switch a DAC's bit currents ON and OFF. The biggest glitches occur at the instant of major code changes in input logic levels. If a natural binary code sequence is examined (say table 1.1), it will be seen that transitions between some adjacent code values require that several bits change at once. All bits changed in the code change at half full scale (that is, 011...1 to 100...0), this is the most major code change. If DAC switches turn bit currents OFF faster than they turn ON, a DAC with its input logic levels incremented through this half scale major code change will have all bits OFF for a brief time. The DAC output will exhibit a large transient error (a large glitch) before it settles to its new value one LSB higher than its previous value. Figure 2.17 shows the type of output which is to be expected.

The size of observed voltage output glitches is markedly influenced by the

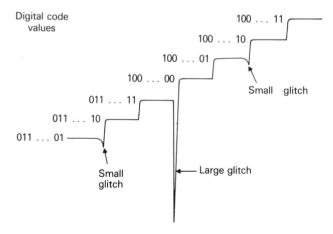

Figure 2.17 Glitch at major code change due to bit currents being turned OFF faster than turned ON

dynamic response characteristics of the output circuitry which is used to change a DAC output current into an output voltage. Very rapid changes in output current occur during a glitch and in many cases the slew rate of the analog output circuitry prevents the output voltage from following these output current changes. This makes the observed output voltage glitch less than the output current glitch which causes it. A small capacitive load connected in parallel with the load resistor driven by a current output DAC can reduce output voltage glitches, but at the expense of an increase in output settling time. In DACs which use an output operational amplifier the amplifier slew rate will normally prevent its output voltage from following very rapid glitch current changes. [1] However, the glitch will tend to prolong output settling time.

In applications in which glitches are a problem the engineer needs to select a DAC design which minimises them. [2] Careful attention to the matching of DAC switching times is a feature of such designs. However, an irreducible minimum will be set by inequalities in the rise and fall response times of the digital logic circuitry which supplies the DAC's digital inputs. High speed logic will reduce the duration of the glitch and hence reduce glitch energy.

Glitches cannot be completely eliminated but they can be prevented from appearing at the output. Linearly filtering out glitches is not satisfactory because the glitches have far from equal magnitudes and they do not occur at equally spaced time intervals. One technique which is commonly used to suppress output glitches is to follow a DAC with a sample/hold device. The sample/hold acts as a deglitcher. The principle underlying the use of a sample/hold as a deglitcher is illustrated in figure 2.18.

The action of the sample/hold is modelled in terms of an ideal switch and a capacitor; readers are referred elsewhere for details of practical sample/holds. [1, 3, 4, 5] The sample/hold shown in figure 2.18 spends most of its time in the sample or track mode (switch closed) with its output equal to the DAC's output. Just before the DAC's digital input is changed to a new value the sample/hold is put in the hold mode (switch open) say by a hold pulse derived from a monostable. After the DAC's output glitch has occurred and the DAC's output has settled the sample/hold is returned to the sample mode. This sequence of operation prevents the glitch from appearing at the output of the sample/hold. Any transient error in the sample/hold output as it acquires its new value is of course still present. The technique inevitably introduces an extra delay due to the acquisition time of the sample/hold—the delay limits the rate at which the DAC output can be updated to new values.

2.9 SELF-ASSESSMENT EXERCISES

1. Which of the following statements are true?
    (i) The bit switches in a DAC are used: (a) to apply logic levels to the DAC; (b) to switch binarily related current increments to the DAC's output.
    (ii) A DAC's bit currents are: (a) components of the DAC's output current

Digital to Analog Converters

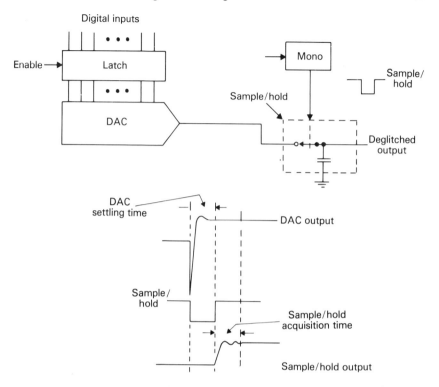

Figure 2.18 Sample/hold acts as deglitcher

determined by the logic states of its digital input bits; (b) current increments which flow in a DAC's input digital bit lines.

(iii) R-$2R$ networks are often used in multi-bit DACs in preference to binarily weighted resistors because: (a) fewer resistors are needed; (b) fewer different resistor values are needed; (c) resistor values do not need to be as close tolerance.

(iv) A basic DAC will accept: (a) both serial and parallel digital input data; (b) serial input data only; (c) parallel input data only.

(v) The main factors influencing the output settling time of a practical DAC are: (a) the dynamic characteristics of the analog output circuitry; (b) the response times of the digital logic circuitry; (c) the type of digital code used; (d) the magnitude of the bit current increments involved in an output change.

(vi) Glitches in a DACs output are due to: (a) capacitive loading at the output; (b) unequal turn on and turn off times of DAC switches; (c) slew rate of the analog circuitry.

Fill in the blanks in the following statements.
(a) A DAC's output current can be converted into a low output impedance

voltage signal by using ....
(b) The output voltage compliance of a current output DAC is. ....
(c) A multiplying DAC is one which uses ....
(d) The output of a multiplying DAC is proportional to the product of . . . .
(e) The internal operational amplifier in a DAC-08 is used to ....
(f) A DAC-08 offers a high input impedance to negative reference voltages because of ....
(g) The direction of the output current flow in a DAC-08 is ....
(h) A unipolar output natural binary coded DAC can be given a bipolar output with sign magnitude code by using an extra sign bit to ....
(i) An offset binary coded DAC can be changed to a two's complement coded DAC by ....
(j) An input digital storage register added to a DAC allows the DAC to . . . .
(k) A combination of .... allows a DAC to accept its input data in serial form.
(l) The settling time of a DAC is. ....
(m) A sample/hold used as a deglitcher limits the rate at which the DAC's output can be updated to new values because of the. .... of the sample/hold.

3. A 10 V reference and binarily weighted resistors are used together with switches to make a 4-bit DAC. If the MSB current increment is 1 mA what resistor values are used? Find the short circuit output current produced by the DAC for the following digital inputs
   (a) 1111
   (b) 1011
   (c) 0011.

4. An R-$2R$ ladder network a 10 V reference and digitally operated switches are used to make a 4-bit natural binary DAC. The DAC produces a short circuit output current of 1.875 mA when all bit switches are ON. Sketch the arrangement giving resistor values.

5. Sketch a circuit showing how you would use an operational amplifier, a +10 V reference and a current switched R-$2R$ network (of the type shown in figure 2.5) to form a DAC which implements the conversion relationship described by the equation

$$V_o = -10\left(x_1 2^{-1} + x_2 2^{-2} + \ldots + x_6 2^{-6}\right)$$

Show how the circuit could be modified using a second operational amplifier configured as a current inverter (see A_2, figure 2.13) so as to produce an output voltage proportional to the difference between the networks normal and complementary current outputs $(I_o - \bar{I}_o)$. Give the equation which represents

Digital to Analog Converters

the conversion relationship implemented by the modified arrangement. What code does it represent?

6. The digital input 10001000 applied to a DAC-08 causes it to produced an output current I_o = 1.0625 mA. Find the value of the DAC reference current and the value of the complementary current output $\bar{I}_o$. Find also the values of I_o and $\bar{I}_o$ for the digital inputs 01111111, and 01110111.

7. The DAC shown in figure 2.11 is to be operated with a reference current 2 mA. What resistor values should be used in the circuit if V_{ref} = 10 V and the nominal full scale analog output range is to be ± 10 V? If the DAC has 6 bits find the output voltage for the following digital input codes
    (a) 000000
    (b) 000101
    (c) 011111
    (d) 100001
    (e) 111111

8. What is the value of the DAC-08 reference current in figure 2.12? Assume resistors have the values indicated in the figure. Find the values of $I_o, \bar{I}_o$ and V_o in figure 2.12, for the following digital input codes
    (a) 00000000
    (b) 00000011
    (c) 10000000
    (d) 10000001
    (e) 11111110

9. The digital inputs of a 10-bit natural binary counter are used to supply the digital inputs to a 10-bit natural binary coded DAC. Sketch the analog output waveform produced by the DAC if the counter is incremented up by a 1 MHz clock signal. Find the frequency of this output wave. Indicate what happens if the counter is made to count down.

10. A DAC with output settling time 500 ns uses a sample/hold module with acquisition time 2 μs as a deglitcher. Find the maximum frequency at which the DAC's output can be updated to new values.

3 Analog to Digital Conversion Techniques

In this chapter some of the more commonly used techniques for implementing an A/D conversion are examined; the discussion is limited to those designs for which, at the time of writing, there are IC devices available. The main techniques to be discussed are conveniently classified under two general headings: feedback A/D converter designs and integrating A/D converter designs.

3.1 FEEDBACK A/D CONVERTERS

The general circuit technique underlying the operation of a feedback A/D converter is illustrated by the block schematic in figure 3.1; the system consists of

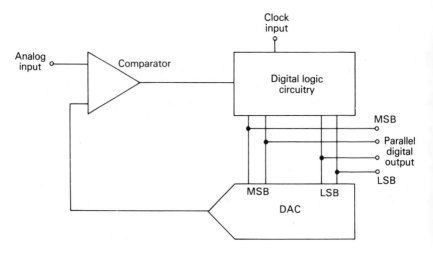

Figure 3.1 Basic schematic for feedback A/D converter

a D/A converter, a comparator and digital logic circuitry. The digital logic circuitry increments up the digital input number applied to the D/A converter and the comparator senses when the analog input signal produced by the D/A

Analog to Digital Conversion Techniques 41

converter becomes equal in value to the analog input signal which is applied to the A/D converter system. Conversion is complete when this equality occurs and the digital number which is then present at the DAC input represents the digitally encoded value of the analog input signal. The ramp-type A/D converter, the tracking A/D converter and the successive approximation A/D converter are feedback designs based on the general schematic of figure 3.1—the three techniques differ in the type of digital logic circuitry which they use.

3.2 RAMP-TYPE A/D CONVERTER

The *ramp-type A/D converter* (count up A/D converter) is probably the simplest in concept; in this design the digital logic circuit consists essentially of a counter. At the start of a conversion the counter is set to zero, it then counts up clock pulses and the digital logic levels representing the count are applied to the logic inputs of the D/A converter. The count is stopped by the comparator when the D/A converter output becomes equal to the externally applied analog input signal and the stored count then constitutes the digital output of the A/D converter system.

A ramp-type A/D converter system can be implemented by simply adding a comparator to the DAC counter system previously investigated (figure 2.9). A suitable arrangement is shown in figure 3.2. The data inputs of the 4-bit binary counter (type 4191) are connected to ground (logical '0') and bringing the load inputs (pin 11) to ground (logical '0') sets the counter to zero. When the load input is returned to logic '1' (open) clock pulses are counted and the DAC output is incremented until the voltage $I_o R_{in}$ becomes equal to the analog input voltage. The comparator output then goes to state '1' and stops the count. The stored count represents the natural binary digitally encoded value of the analog input signal expressed as a fraction of the full scale analog input where full scale analog input has the value $I_{ref} R$. (Note $V_{in} = (255/256) I_{ref} R_{in}$ gives digital output 11111111.)

In a ramp-type A/D converter the conversion is completed at the instant at which the DAC's analog output becomes equal to the analog input signal. The system in figure 3.2 uses a current comparison technique and in this case conversion is completed when $I_o = V_{in}/R_{in}$. If the analog input now decreases, the digital output remains constant or 'holds'; if the analog input increases the counter increments up again until equality of analog input and DAC output is again reached. The digital output in a ramp-type ADC thus represents the maximum value of the analog input during the time between counter resets.

The conversion time in a ramp-type A/D converter is not fixed but depends on the size of the analog input expressed as a fraction of the full scale. In the system of figure 3.2

$$\text{conversion time} = \frac{V_{in}}{I_{ref} R_{in}} \times 2^n T_c \qquad (3.1)$$

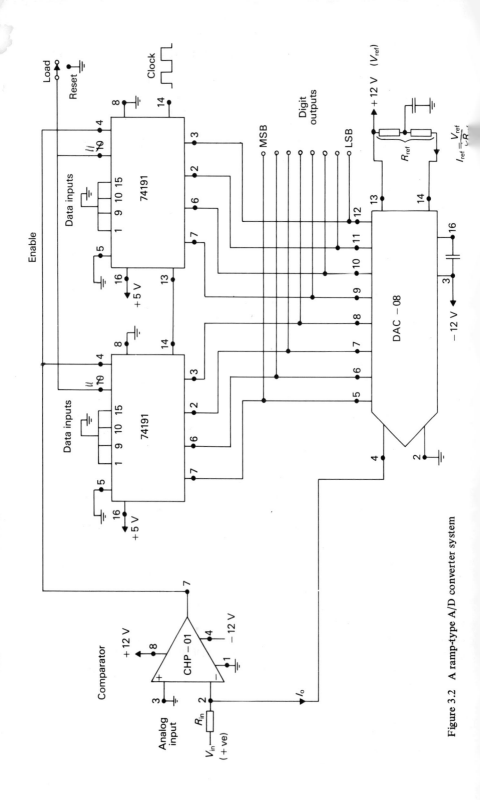

Figure 3.2 A ramp-type A/D converter system

Analog to Digital Conversion Techniques

where n is the number of logic bits in the DAC, ($n = 8$ in figure 3.2) and T_c is the period of the clock pulses. For example if the clock frequency where 1 MHz, $T_c = 1$ μs and a full scale less one LSB conversion involving all 8 bits would take $(255/256) \times 256 = 255$ μs.

In any digital representation of an analog variable there is always a possible error or uncertainty (quantisation uncertainty, section 1.3). An analog variable can take on a continuum of values whereas a discontinuous variation is inherent in a numerical or digital representation of analog quantity. The ramp-type converter system of figure 3.2 seeks to establish equality between the analog input voltage V_{in} and the voltage $I_o R_{in}$ developed by the DAC. V_{in} can vary continuously but I_o, the DAC output current varies in steps, one LSB current increment, $I_{ref}/2^n$, at a time. The system cannot therefore establish the exact equality $V_o = I_o R_{in}$. The system stops the DAC's output from incrementing up when $I_o R_{in}$ first exceeds V_{in}. In so doing it finds the digital bit values which satisfy the relationship

$$\left(\frac{V_{in}}{I_{ref} R_{in}} + \frac{1}{2^n}\right) > \left(x_1 2^{-1} + x_2 2^{-2} + \ldots + x_n 2^{-n}\right) > \frac{V_{in}}{I_{ref} R_{in}} \quad (3.2)$$

$n = 8$ for DAC –08.

All analog values in the range one LSB, $(I_{ref} R_{in})/256$, below the analog value which satisfies the exact equality $V_{in} = I_o R_{in}$ give rise to the digital code which produces that equality. If it is required that a $\pm\frac{1}{2}$ LSB quantisation uncertainty be symmetrical about nominal code values (as described in section 1.3) a $\frac{1}{2}$ LSB current increment must in effect be added to the DAC's output current.

The comparator in figure 3.2 may be thought of as analogous to an operational amplifier with shunt negative feedback. [1] The digital logic circuitry and DAC complete a feedback loop which seeks to maintain the comparator output terminals at the same potential. (In the same way that an operational amplifier with negative feedback seeks to maintain its differential input terminals at the same potential.) The comparator input (pin 2 in figure 3.2) may be compared to an operational amplifier virtual ground or summing point. It is a simple matter to add a current offset at the comparator summing point. A resistor with value such that $\bar{V}_s/R = I_{ref}/2^{N+1}$ connected between the comparator summing point and the negative supply $\bar{V}_s$, can be used to set the systems $\pm\frac{1}{2}$ LSB quantisation range.

3.3 TRACKING A/D CONVERTERS

Tracking converters are similar to ramp-type converters but their digital logic circuitry consists of an up/down counter instead of an up counter. Some outline examples of tracking converter configurations are given in figure 3.3. Working systems can easily be assembled by making minor interconnector changes between the devices shown in figure 3.2.

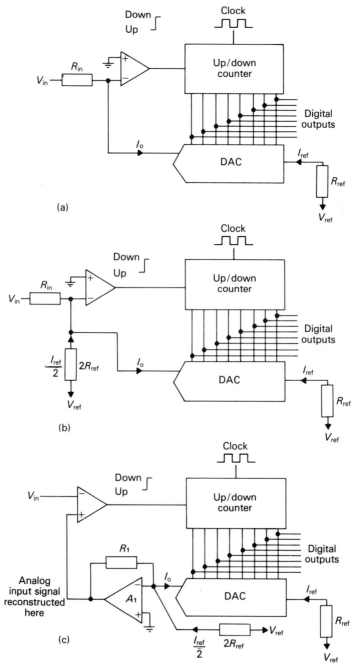

Figure 3.3 Tracking A/D converter configurations: (a) unipolar tracking converter (current comparison); (b) bipolar tracking converter (current comparison); (c) high input impedance bipolar tracking converter (voltage comparison)

Analog to Digital Conversion Techniques

In the systems shown in figure 3.3 the comparator output controls the counting mode (up or down count) so as to make the DAC's output current increment in such a way as to bring the comparator input terminals to the same potential. A current comparison is performed by the system in figure 3.3a. If $V_{in}/R_{in} > I_o$ the inverting input of the comparator is made positive with respect to its non-inverting input. The comparator output thus goes low, causing the counter to increment up the binary digital number applied to the DAC. If $V_{in}/R_{in} < I_o$ the comparator output goes high and the DAC's output is incremented down. With a constant analog input signal, once I_o gets within one LSB of V_{in}/R_{in}, the system gives a digital output which alternates or dithers between the code values spanning the theoretically 'correct' code values.

A current comparison is also performed by the system in figure 3.3b. An offset current $I_{ref}/2$ is added to the comparator summing point. The offset gives the system a bipolar input capability with an offset binary code relating analog input and output digital code values.

The system shown in figure 3.3c uses a voltage comparison technique. An operational amplifier, A_1, is used to convert the DAC's output current to an output voltage. The action of the system is to seek to establish equality between this output voltage and the analog input signal. The comparator and associated feedback loop is analogous to an operational amplifier used in the series feedback configurations, (the follower configuration). The voltage comparison system has the high input impedance which is a characteristic of series feedback circuits. [1] The offset current used to provide bipolar operation is added at the operational amplifier summing point.

3.3.1 Slew Rate Limitations

If a varying analog input signal is applied to a tracking converter system, the action of the feedback loop forces the digital signal which is generated to track the analog signal variations. The digital signal then continuously reflects the value of the changing analog input signal. There is, however, a maximum rate of change of the analog input signal for which this tracking condition is maintained.

The DAC output is incremented one LSB at the occurrence of each clock pulse. The maximum rate at which the DAC output can change is thus controlled by the clock frequency. If the analog input rate of change exceeds this maximum rate the digital output is no longer an accurate representation of the analog signal. The maximum rate of change is called the *loop slew rate*.

$$\text{slew rate} = \frac{V_{fs}}{2^n} \times f_{cp} \qquad (3.3)$$

where f_{cp} = clock frequency, V_{fs} = the full scale analog input signal range (full scale negative to full scale positive in the case of a bipolar converter), n = number of bits and $V_{fs}/2^n$ represents the analog magnitude of the LSB. In the systems shown in figures 3.3a and 3.3b $V_{fs} = I_{ref}R_{in}$. In figure 3.3c $V_{fs} = I_{ref}R_1$.

The slew rate limitation of a tracking converter sets an upper frequency limit to the full scale sinusoidal signal that can be accurately digitised.

A full scale sinusoid may be represented by the expression

$$v = \frac{V_{fs}}{2} \sin 2\pi f t$$

The maximum rate of change of this sinusoid is obtained by differentiation as

$$\left| \frac{dv}{dt} \right|_{max} = 2\pi f \frac{V_{fs}}{2}$$

The maximum frequency for which the slew rate is not exceeded is thus determined by

$$2\pi f_{max} \frac{V_{fs}}{2} = f_{cp} \frac{V_{fs}}{2^n}$$

or

$$f_{max} = \frac{f_{cp}}{2^n \pi} \qquad (3.4)$$

3.3.2 Tracking Converter Acts as Sample/Hold

Sample/hold modules are used to obtain instantaneous values or samples of changing analog signals. A sample hold is an input/output device which has two operating modes: sample and hold. In the sample mode of operation output signal is equal to input signal and tracks or follows input signal time variations. On the receipt of a command to hold the output signal is ideally held constant at a value equal to that of the input signal at the instant the hold command was received.

A tracking converter system with its counter operational and supplied with a continuous train of clock pulses may be thought of as a sample/hold in the sample mode. The system's digital out put tracks or follows input signal variations. In a system using the voltage comparison technique (figure 3.3c) a reconstructed version of the analog input signal is continuously available.

A tracking converter is put into the hold mode by simply stopping its counter from incrementing. The stored count then represents the value of the analog input signal at the instant that the counter was disabled. The value of the stored count and the reconstructed analog input can be held indefinitely without loss. In this respect the system has an advantage over an analog sample/hold module. Analog sample/holds do not have an indefinitely long hold time capability. The output signal of an analog sample/hold slowly decays in the hold mode.

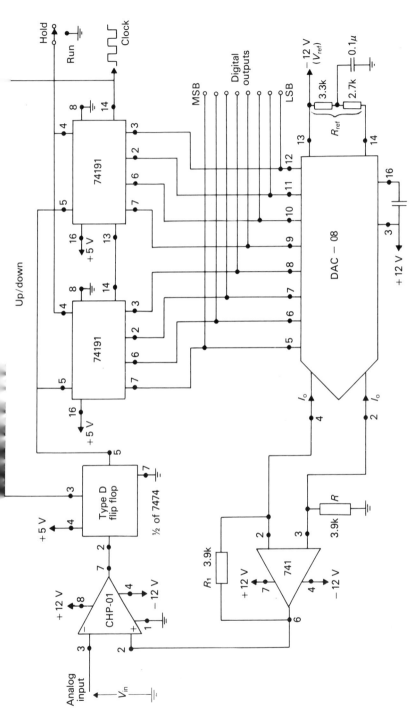

Figure 3.4 A bipolar input tracking A/D converter system

3.3.3 A Practical Tracking Converter System

Detailed pin connections for a tracking converter system are shown in figure 3.4. The circuit is intended to provide the reader with a convenient basis for experimental investigation. A type D flip-flop is shown connected between the comparator output and the counter's up/down mode control input. This ensures that the comparator completes an output transition before the next change in counting mode occurs. The flip-flop can be omitted if high clock frequencies are not used.

The operational amplifier shown in figure 3.4 produces a bipolar output signal $(I_o - \bar{I}_o) R_1$ which depends upon the difference between the DAC-08's normal and complementary current outputs. This gives the system a bipolar symmetrical offset binary code. The analog input signal is reconstructed at the output terminal of the operational amplifier, the loop forces the operational amplifier output to track analog input signal variations. The oscilloscope traces reproduced in figure 3.5 show the effect on this reconstructed output signal of

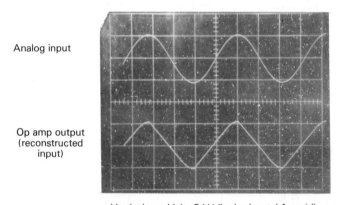

Vertical sensitivity 5 V/div; horizontal 1 ms/div

Figure 3.5 Waveforms obtained from circuit of figure 3.4 when input signal exceeds loop slew rate

using an analog input signal whose rate of change exceeds the loop slew rate. The loop slew rate is (see equation 3.3)

$$\text{loop slew rate} = \frac{2 I_{ref} R_1}{2^n} f_{cp}$$

where

$$I_{ref} = \frac{12 \text{ V}}{6 \text{ k}\Omega} = 2 \text{ mA}$$

Analog to Digital Conversion Techniques

A clock frequency 100 kHz was used, so substituting this and other parameter values gives

$$\text{loop slew rate} = \frac{2 \times 3.9}{128} \times 10^5$$

$$= 6.09 \times 10^3 \text{ V/s}$$

Examination of the slew rate limited portion of the reconstructed signal in figure 3.5 gives the measured loop slew rate as

$$\frac{9.5}{1.6 \times 10^{-3}} = 5.9 \times 10^3 \text{ V/s}$$

3.4 SUCCESSIVE APPROXIMATION A/D CONVERTER

The *successive approximation* A/D conversion technique provides a more rapid conversion than the other two feedback techniques. The digital logic circuitry used in this technique instead of incrementing the D/A converter output one LSB at a time, performs a series of 'trial' conversions. In the first trial the control logic applies the MSB to the D/A converter and the analog output of the D/A converter ($\frac{1}{2}$ full scale) is compared with the analog input signal by the comparator. If the DAC's output is less than the analog input the MSB is retained; if the DAC's output is greater than the analog input the MSB is switched OFF. The control logic then goes on to apply the next MSB which is retained or discarded depending on the result of the comparison between the DAC's output and analog input. The process of trying the addition of successively smaller bits and retaining or discarding them depending on the result of the trial, goes on until the LSB is reached; the conversion is then complete.

A simplified representation of the timing sequence which occurs in a typical 4-bet successive approximation A/D conversion is shown in figure 3.6; the analog input is assumed to lie between 9/16 and 10/16 full scale. On the clock pulse low to high transition, at the start of the conversion, all bits except bit 1 are set to zero. The analog output of the DAC produced by bit 1 is $\frac{1}{2}$ full scale; this is less than the analog input signal—the comparator indicates this by giving a logical '1' which is fixed into the bit 1 register on the next low to high clock pulse transition, at the same time bit 2 is switched on. The DAC output of 12/16 full scale is now bigger than the analog input signal, and the comparator sets a logical 0 in the bit 2 register on the next clock low to high transition at which time bit 3 is switched ON. So the conversion proceeds and is completed (digital output 1001) on the fifth low to high transition of the clock pulse.

The digital logic circuitry required to implement a successive approximation A/D conversion can be assembled using standard TTL logic gates and flip flops, but from a user standpoint it is generally more convenient to make use of an MSI device called a successive approximation register (SAR).

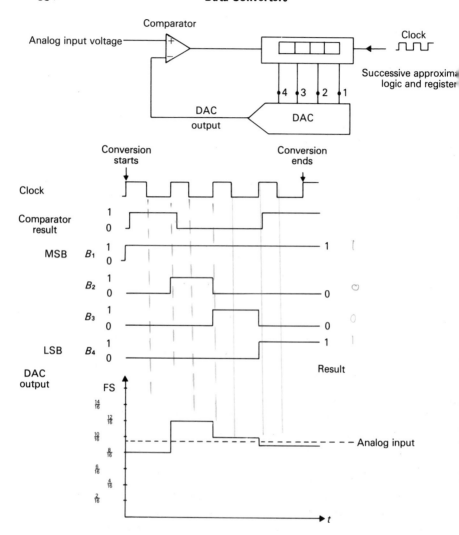

Figure 3.6 Timing sequence in a 4-bit successive aprroximation A/D conversion. Note: If DAC output is greater than analog input the analog input comparator gives a logical '0' which is entered and retained on next low to high of clock pulse

3.4.1. A Successive Approximation Register

Inspection of figure 3.6 shows that in a successive approximation A/D converter the digital control circuitry (the successive approximation register) is required to perform a serial to parallel digital data conversion. A successive approximation register is in effect a special purpose serial to parallel data converter. The serial data is that which appears at the comparator output in the sequence of trials which constitute the successive approximation conversion. The comparator

Analog to Digital Conversion Techniques

output indicates the result of each trial and at the same time gives one of the required bit values. The comparator output in effect gives a serial representation of the digital word which is generated by the conversion sequence. Starting with the MSB the successive approximation logic sets these serial bit values into their appropriate register positions where they remain until a new conversion is initiated. Also as each bit value is found and fixed into its register, lower order bits are set at the values appropriate for the next trial.

Figure 3.7 shows the pin connections for an MSI TTL successive approximation register, the 2502 device. The 2502 has 11 output pins, 3 input pins,

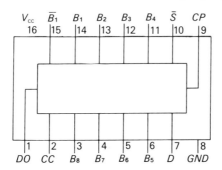

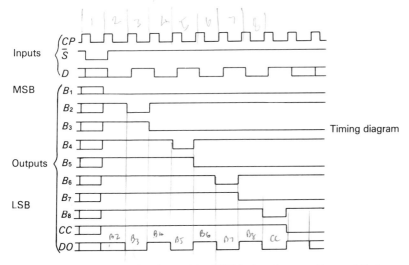

Figure 3.7 Successive approximation register 2502 device pin connections and timing diagram

power supply positive (+5 V) and ground pins—it is a 16-pin dual in line device. The output pins are: the bit registers, B_1 through to $\bar{B}_8$ ($\bar{B}_1$ is also available for two's complement coding); a serial data output DO, an output which signals conversion complete, CC. The inputs are: CP, a clock signal must be applied here it controls the timing of operations; $\bar{S}$, an input is applied here to initiate a conversion sequence; D, the serial input data to the register is applied here.

The functional operation performed by the 2502 register is summarised by the timing diagram given in figure 3.7. A conversion sequence is initiated by bringing the $\bar{S}$ input low. The following transitions then occur on the first low to high transition of the clock pulse, B_1 register is set low all other registers are set high, CC goes high indicating that a conversion is being performed. (It should be noted that this initial register set up is the complement of that given in figure 3.5). After the first clock low to high transition has occurred, $\bar{S}$ must be brought back high before the next clock pulse.

In figure 3.7 the data input on the D line is shown changing immediately after a clock low to high transition. The data shown is an alternating sequence of low and high levels. When the SAR is used as part of a successive approximation A/D converter the data input is supplied by the system's comparator. In such a system the actual data sequence applied at the D input depends on the size of the analog signal which is being converted.

On the occurrence of the second low to high clock transition the data present at the D input (as shown in figure 3.7) is set permanently in the B_1 register for the duration of the conversion sequence. At the same time the B_2 register is set low and all lower order registers are set high. On the third clock transition the data present at the D input is set permanently into the B_2 register, the B_3 register is set low and all lower order registers are set high. The sequence continues until the ninth low to high clock transition when the final data bit is set into the B_8 register and the CC output goes low indicating that the conversion has been completed. The data then present at the register remains unchanged by further clock pulses until a new conversion sequence is initiated by bringing $\bar{S}$ low.

3.5 PRACTICAL SUCCESSIVE APPROXIMATION A/D CONVERTER SYSTEMS

A wide range of ready built successive approximation A/D converters are commercially available in modular form and in hybrid and monolithic integrated circuit form. Although specific converters may differ in the fine details of their performance they are all very similar in their functional mode of operation. The engineer must have a thorough understanding of this functional operation if he is to sort his way successfully through the maze of commercially available converters in order to choose one which best suits his application.

Understanding is reinforced as a result of experimentation with a specific converter system. Examples of practical systems which can be assembled using inexpensive IC functional elements are given in this section. As shown, the systems are intended for an experimental breadboard evaluation but they can equally serve as a basis for a converter system to be used in a practical application.

Analog to Digital Conversion Techniques

A basic unipolar 8-bit successive approximation A/D converter system is shown in figure 3.8. The system uses an 8-bit DAC, the DAC-08, an 8-bit SAR,

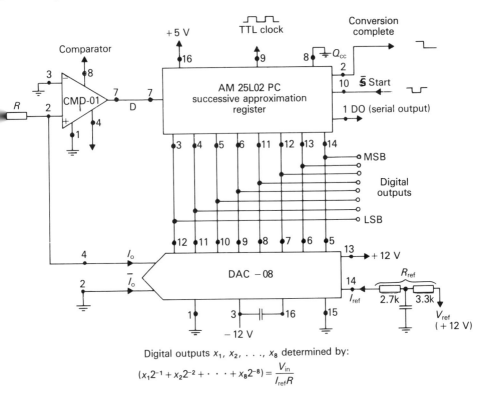

Figure 3.8 Successive approximation A/D converter system

Digital outputs $x_1, x_2, \ldots, x_8$ determined by:

$$(x_1 2^{-1} + x_2 2^{-2} + \cdots + x_8 2^{-8}) = \frac{V_{in}}{I_{ref} R}$$

the 2502 device, and a comparator (CMP-01). Positive and negative supplies are used to power the DAC and comparator, and for simplicity the DAC reference is provided by the +12 V supply. In a practical application of the system a separate reference voltage supply would normally be used for greater accuracy and stability.

The system's operation can be investigated by applying a repetitive clock pulse (frequency, say, 10 kHz) and using an oscilloscope to measure the signals which appear at various points in the system. The conversion complete signal (at pin 2 of the SAR) is connected to the start conversion input (pin 10 of the SAR) for repetitive conversions. If the signal which appears at *CC* is used to trigger the oscilloscope externally, a functional timing diagram (similar to that shown in figure 3.7) can be obtained by transferring the oscilloscope input leads in sequence to the various measurement points. As an alternative, clock pulses can be applied manually, one at a time, and logic indicators used to observe signal states after each clock pulse. The data sequence appearing at

D during a conversion depends on the value of the analog input signal which is being converted. This analog input should be held constant during a conversion.

The DAC produces an output current I_o which is related to its input logic levels by the relationship

$$I_o = I_{ref}(x_1 2^{-1} + x_2 2^{-2} + \ldots + x_n 2^{-n})$$

$x_1, x_2, \ldots, x_n$ are the logic levels at the SAR bit registers B_1, B_2 etc.; $n = 8$ for the 8-bit DAC

$$I_{ref} = \frac{V_{ref}}{R_{ref}}$$

and in figure 3.8

$$I_{ref} = \frac{12\,V}{6\,k\Omega} = 2\,mA$$

In each trial the comparator finds out whether V_{in} is greater or less than $I_o R_{in}$. The comparator output goes high (a '1') if $V_{in} > I_o R_{in}$ and goes low if $V_{in} < I_o R_{in}$.

The sequence of trials performed in a conversion is designed to find the bit values $x_1, x_2, \ldots$, etc., which establish equality between V_{in} and $I_o R_{in}$. It is in the nature of an A/D conversion that an exact equality cannot be attained (quantisation uncertainity, see section 1.3). The system shown in figure 3.8 find the bit values which satisfy the relationship

$$\frac{V_{in}}{I_{ref} R_{in}} + \frac{1}{2^n} > \left(x_1 2^{-1} + x_2 2^{-2} + \ldots + x_n 2^{-n}\right) > \frac{V_{in}}{I_{ref} R_{in}} \tag{3.5}$$

$n = 8$ for the 8-bit system.

All analog values which lie in the range $(1/256) I_{ref} R_{in}$ below the analog value satisfying the exact equality $V_{in} = I_o R_{in}$ give rise to the digital code value which produces that exact equality.

The sequence involved in a conversion may be understood if we consider what happens when a specific value of the analog input signal V_{in} is applied, say $V_{in} = 6\,V$. We express 6 V as

$$\frac{6}{I_{ref} R_{in}} \times 256 = \frac{6}{2 \times 5} \times 256 = 153.6 \quad 1/256 \text{ ths of } I_{ref} R_{in}$$

We note that the comparator in figure 3.8 will give a low output (state 'O') if $I_o/I_{ref} > 153.6/256$ and a high output (state '1') if $I_o/I_{ref} < 153.6/256$. The sequence that occurs may then be summarised as shown in table 2.1.

The comparator result obtained in the eighth trial is fixed into the B_8 register on the ninth clock transition. The data present in the bit registers is then a valid

Analog to Digital Conversion Techniques

Table 2.1

Trial Number (clock pulse no)	SAR Registers MSB x_1	x_2	x_3	x_4	x_5	x_6	x_7	LSB x_8	$\dfrac{I_o}{I_{ref}}$	Comparator Result
1	0	1	1	1	1	1	1	1	$\frac{127}{256}$	1
2	1	0	1	1	1	1	1	1	$\frac{191}{256}$	0
3	1	0	0	1	1	1	1	1	$\frac{159}{256}$	0
4	1	0	0	0	1	1	1	1	$\frac{143}{256}$	1
5	1	0	0	1	0	1	1	1	$\frac{151}{256}$	1
6	1	0	0	1	1	0	1	1	$\frac{155}{256}$	0
7	1	0	0	1	1	0	0	1	$\frac{153}{256}$	1
8	1	0	0	1	1	0	1	0	$\frac{154}{256}$	0
Conversion completed 9th clock pulse	1	0	0	1	1	0	1	0		

representation of the analog input and the conversion is completed. Note that as the conversion proceeds, the data appears in serial form at the *DO* output of the SAR.

A graphical presentation of the digital output/analog input relationship which is implemented by the system is shown, in part, in figure 3.9a. The $\pm\frac{1}{2}$LSB quantisation uncertainty is not centred about nominal code values. The system requires a slight modification if it is to be made to implement a conversion relationship of the form shown in figure 3.9b. The modification must add an offset current ($\frac{1}{2} I_{ref}/2^n$) = ($\frac{1}{2} I_{ref}/256$) ($\frac{1}{2}$ LSB) to the DAC's output current to give bit values which satisfy the relationship

$$\left(\frac{V_{in}}{I_{ref}R_{in}} + \frac{1}{2^n}\right) > (x_1 2^{-1} + x_2 2^{-2} + \ldots + x_n 2^{-n}) + \frac{1}{2} \times \frac{1}{2^n}$$

$$> \frac{V_{in}}{I_{ref}R_{in}}$$

which can be expressed as

$$\left(\frac{V_{in}}{I_{ref}R_{in}} - \frac{1}{2^{n+1}}\right) > (x_1 2^{-1} + x_2 2^{-2} + \ldots + x_n 2^{-n}) >$$

$$> \left(\frac{V_{in}}{I_{ref}R_{in}} + \frac{1}{2^{n+1}}\right) \quad (3.6)$$

Equation 3.6 represents an A/D conversion relationship in which the $\pm\frac{1}{2}$ LSB is

Data Converters

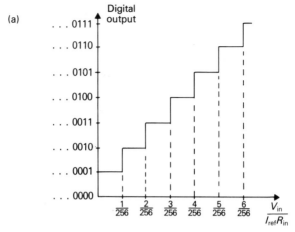

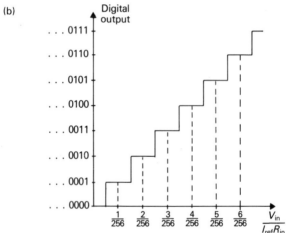

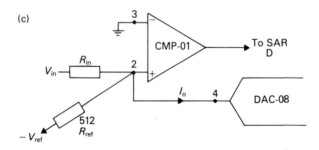

Figure 3.9 (a) Conversion relationsjip implemented by figure 3.8; (b) modified relationship with $\pm \frac{1}{2}$ LSB uncertainty centred; (c) modification required to figure 3.8 to centre $\pm$ LSB quantisation uncertainty

Analog to Digital Conversion Techniques

centred about nominal code values as shown in figure 3.9b. The circuit modification required in figure 3.8 consists of connecting a resistor value $2^{n+1} R_{ref}$ between the non-inverting comparator input (pin 2) and a negative reference voltage supply – V_{ref} as shown in figure 3.9c.

An alternative arrangement of the functional elements in figure 3.8 is possible; it is shown in figure 3.10. The alternative arrangement uses the DAC-08's $\bar{I}_o$

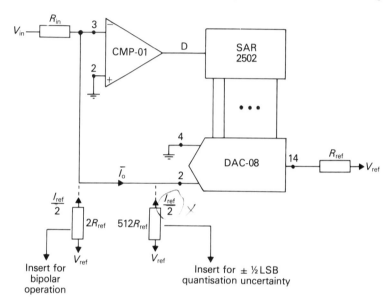

Figure 3.10 Alternative arrangement of functional elements gives succesive approximation converter a complementary binary code

current output line. The increments of the $\bar{I}_o$ current are switched on by low logic levels applied to the DAC's digital inputs; this requires that the input signal and $\bar{I}_o$ be connected to the inverting input of the comparator as shown. The comparator now gives a low logic level output when $V_{in} > \bar{I}_o R_{in}$ and a high level when $V_{in} < \bar{I}_o R_{in}$. Low levels set in the SAR bit registers switch on the appropriate $\bar{I}_o$ bit current increments.

The arrangement in figure 3.10 gives a complementary binary conversion code. It can, however, be interpreted as natural binary if a low logic level is used to signify binary '1'. Assuming the more usual positive logic convention the x_1 bit values which the system finds in a conversion and applies to the DAC are those which satisfy the relationship

$$\frac{V_{in}}{I_{ref} R_{in}} > (\bar{x}_1 2^{-1} + x_2 2^{-2} + \ldots + \bar{x}_n 2^{-n}) > \left(\frac{V_{in}}{I_{ref} R_{in}} + \frac{1}{2^n} \right)$$

(3.7)

Note that in the conversion code given by equation 3.7 all analog values which lie in the analog range $(1/2^n) I_{ref} R_{in}$ above the nominal analog value corresponding to a code word give rise to that same code value.

The placing of the quantisation uncertainty is different to that given by equation 3.5. The difference arises because of the way in which the 2502 SAR operates. In finding a bit value the SAR first sets that bit low and all lower order bits high. The system in figure 3.10 can be made to give its $\pm \frac{1}{2}$ LSB quantisation uncertainty either side of nominal code values if a current ($\frac{1}{2} I_{ref}/2^n$) ($\frac{1}{2}$ LSB) is subtracted from the DAC's $\bar{I}_o$ output current. Equation 3.7 then becomes

$$\frac{V_{in}}{I_{ref}R_{in}} > (\bar{x}_1 2^{-1} + \ldots + \bar{x}_n 2^{-n}) - \frac{1}{2} \times \frac{1}{2^n} > \left(\frac{V_{in}}{I_{ref}R_{in}} - \frac{1}{2^n}\right)$$

which can be expressed in the same form as equation 3.6, namely

$$\left(\frac{V_{in}}{I_{ref}R_{in}} + \frac{1}{2^{n+1}}\right) > (\bar{x}_1 2^{-1} + \bar{x}_2 2^{-2} + \ldots + \bar{x}_n 2^{-n}) >$$

$$> \left(\frac{V_{in}}{I_{ref}R_{in}} - \frac{1}{2^{n+1}}\right) \qquad (3.8)$$

Subtraction of the $\frac{1}{2}$ LSB current increment from the DAC's $\bar{I}_o$ output current can be accomplished by connecting a resistor value $2^{n+1} R_{ref} = 512 R_{ref}$ between the DAC's output and the positive reference voltage which is used to supply the DAC's reference current. The connection is indicated in figure 3.10.

3.5.1 Bipolar Successive Approximation Converter Systems

Unipolar natural binary coded successive approximation converter systems can usually be easily modified to accept bipolar analog input signals. The modification involves offsetting the DAC's output by an amount equal to half its full scale range and it gives the system an offset binary code.

The systems shown in figures 3.8 and 3.10 are changed into bipolar systems by connecting a resistor value $2R_{ref}$ between the DAC's output and the DAC's positive reference voltage V_{ref}. The modification in effect subtracts a current $I_{ref}/2$ from the current which the DAC takes from the comparator's input. With the DAC's output offset $I_{ref}/512 - I_{ref}/2$ the system in figure 3.8 implements a conversion in which code values are governed by the relationship

$$\left(\frac{V_{in}}{I_{ref}R_{in}} + \frac{1}{2^{n+1}}\right) > (x_1 2^{-1} + x_2 2^{-2} + \ldots + x_n 2^{-n}) - \frac{1}{2} >$$

$$> \left(\frac{V_{in}}{I_{ref}R_{in}} - \frac{1}{2^{n+1}}\right) \qquad (3.9)$$

Analog to Digital Conversion Techniques 59

where $n = 8$. With $I_{ref} = 2$ mA and $R_{in} = 5$ kΩ, analog full scale negative is $(-5 \pm 10/512)$ V corresponding to the digital code 00000000. Analog zero is $(0 \pm 10/512)$ V with digital code 10000000. Analog full scale positive is $[5 \times (254/256) \pm (10/512)]$ V with digital code 11111111. The code is offset binary. With the DAC's output offset by $-(I_{ref}/512 + I_{ref}/2)$ the system in figure 3.10 implements a conversion code in which code values are governed by the relationship

$$\left(\frac{V_{in}}{I_{ref}R_{in}} + \frac{1}{2^{n+1}}\right) > (\bar{x}_1 2^{-1} + x_2 2^{-2} + \ldots + \bar{x}_n 2^{-n}) - \tfrac{1}{2} >$$

$$> \left(\frac{V_{in}}{I_{ref}R_{in}} - \frac{1}{2^{n+1}}\right) \quad (3.10)$$

where $n = 8$. In this case the digital code for full scale negative is 11111111, the code for zero is 01111111 and the code for full scale positive is 00000000. The code is complementary offset binary.

3.5.2 Two's Complement Bipolar Code Conversions

A two's complement bipolar code is an offset binary code with its MSB (its sign bit) complemented. The 2502 SAR provides the complement of the bit stored in register B_1 at the device pin 15 ($\bar{B}_1$). The bipolar offset forms of the system in figures 3.8 and 3.10 can be used to provide a two's complement code. Connections between the SAR and the DAC are exactly as shown in the diagrams but the MSB of the parallel digital output word is simply taken from SAR pin 15 ($\bar{B}_1$) instead of pin 14. The system in figure 3.8 then provides a normal two's complement code; the system in figure 3.10 provides a complementary two's complement code.

3.6 CONVERSION TIME OF A SUCCESSIVE APPROXIMATION A/D CONVERTER

The time taken by a successive approximation converter to complete a conversion depends on the number of bits involved in the conversion. A repetitive clock pulse controls the timing of the sequence of operations involved in a conversion. The system takes 1 clock period to find each bit value. In an n-bit conversion the value of the nth bit is found on the occurrence of the nth clock pulse after the start of the conversion but its value is not fixed into the nth register until the occurrence of the $(n + 1)$th clock pulse. An n-bit conversion is thus completed in $n + 1$ clock periods. Conversion time T_c is

$$T_c = (n + 1) T_{cp} = \frac{n + 1}{f_{cp}} \quad (3.11)$$

where $f_{cp} = 1/T_{cp}$ is the clock frequency. With a fixed number of bits, conversion time varies inversely with the system's clock frequency. There is a limit to the maximum clock frequency which can be used with a particular system. The limit is set by the various time delays associated with the operation of the system elements.

System delays include the comparator response time (the time taken by the comparator output to change state), the DAC settling time and propagation delays of the digital logic circuitry. Other delays can be introduced as a result of stray impedances arising from the system layout.

System layout is an important factor in high speed A/D converter systems. Some circuit board layout rules [6] are as follows.

(1) Digital ground must be separated from analog ground; they should meet at only one common point.
(2) Digital traces should not cross or be routed near sensitive analog areas. This is particularly important near the sum mode (the comparator input).
(3) The trace from the DAC's output to the comparator's input should be short and it should be guarded by analog ground.
(4) Input analog traces should be short.
(5) The comparator's output should be routed away from its input to minimise capacitive coupling and possible oscillations.

3.6.1 Shortening Conversion Time by Reducing the Number of Bits

In converter applications not requiring the full resolution provided by all device bits, conversion time can be shortened by reducing the number of bits actually used. The successive approximation converter systems given in figures 3.8 and 3.10 can be easily modified to operate with a reduced number of bits. For an n-bit conversion ($n < 8$) the conversion complete signal is simply taken from the $n + 1$ bit register. The signal is valid for one clock period only since the 2502 continues to step through the remaining bits.

A continuous conversion application with n bits is implemented by using the high to low transition of the $(n + 1)$ bit as the successive approximation registers start conversion signal. A possible lock-out condition which could occur on first applying power is avoided by arranging that either the register conversion complete signal or the $(n + 1)$ bit can initiate a new sequence. A suitable arrangement is shown in figure 3.11. Connections shown are for a 4-bit conversion; the register bit 5 going low initiates each new conversion cycle.

It should be remembered that reducing the number of conversion bits increases the quantisation uncertainty. The value of the resistor used to provide the $\frac{1}{2}$ LSB offset must be selected according to the LSB actually used. DAC digital inputs not used in a reduced resolution application should be connected to the logic level which turns their corresponding output bit current increments OFF.

Analog to Digital Conversion Techniques 61

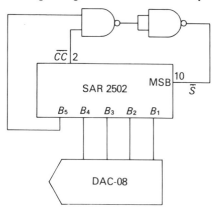

Figure 3.11 Reducing the number of bits in a successive approximation converter (truncating the register)

3.7 ADDING AN OUTPUT DIGITAL STORAGE REGISTER TO A SUCCESSIVE APPROXIMATION CONVERTER

The parallel digital data which is present at the output of a successive approximation converter is a valid representation of the analog input signal only when a conversion has been completed. While a conversion is in progress the digital data is changing. In order that valid digital data should be continuously available, successive approximation converters often use an output digital storage register or output buffer. This output store is updated with new data as it becomes available at the end of a conversion cycle. The register stores the result of a previous conversion while the next conversion cycle takes place.

There are a variety of integrated circuit registers available which can be used to perform as an output storage register. Readers should consult digital device catalogues for details. Many of the newer integrated circuit successive approximation converters which are now available include an output digital register as an integral part of their circuit package. Output registers are normally provided with tri-state outputs to allow permanent connection to a shared data bus (see section 5.10).

Readers may wish to add a digital output buffer to the experimental converter systems of figures 3.8 and 3.10. One way of doing this is shown in figure 3.12. Pin connections for an 8-bit storage latch TTL type 74LS363 are indicated.

The conversion system's conversion complete signal going low is used to trigger a monostable. The monostable generates a single positive going pulse which acts as the enable signal for the storage latch. The latch is transparent when its enable input is high, data present at the latch inputs then appears at the outputs. The data present at the outputs is unaffected by input data changes

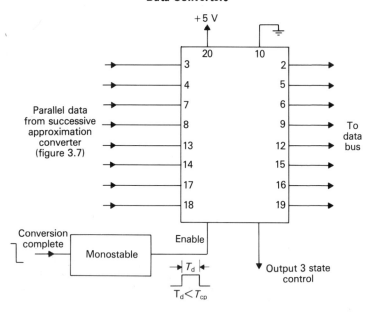

Figure 3.12 Adding an output data latch to a successive approximation converter

when the enable input goes low. The 74LS363 device has 3-state bus-driving outputs. Outputs are inactive in their high impedance state when the logic level applied to the device output control (pin 1) is high.

3.8 INTEGRATING TECHNIQUES FOR A/D CONVERSION

The A/D conversion techniques described in the previous sections give a digital output which depends on the value of the analog input signal which exists at some precise instant in time, and their output is thus affected by the presence of noise on the analog input signal. A variety of A/D conversion techniques have been developed in which the digital output depends on the integral or average value of the analog input signal during some prescribed time interval. Integrating techniques have the advantage of giving repeatable results even in the presence of high frequency noise on the analog input signal. The effect of high frequency noise is averaged out provided that the noise frequencies present are such that

$$\frac{1}{f_n} < T_i$$

where T_i is the integration period.

The noise rejection properties of integrating A/D converters can be put to good effect in making the output of an integrating converter insensitive to any mains interference signal which might be present on the converters input signal.

Analog to Digital Conversion Techniques 63

In order to achieve rejection of mains interference it is simply necessary to make the converter's input signal integrate period a multiple of the period of the a.c. supply waveform. The integral of a sinusoidal signal is zero if the integration is performed over a time interval which is a whole number of the time periods of the sinusoid (see exercise 12).

Monolithic IC devices suitable for the implementation of integrating A/D converters are available from several manufacturers. They are comparatively inexpensive yet nevertheless can provide very accurate conversion. Their main disadvantage, when compared with the techniques discussed previously, is their much longer conversion time.

Currently available integrating converters fall into two categories: those which use the so-called dual-slope technique and those which use the quantised-feedback technique to perform a conversion. The dual-slope technique is the simplest in concept and implementation—it has been the preferred integrating conversion method for a number of years. Quantised-feedback converters are comparative newcomers; however, in some applications they provide several advantages over the older dual-slope techniques.

3.9 DUAL-SLOPE A/D CONVERSION

The operating principles involved in a basic dual-slope A/D conversion may be understood by reference to the simplified circuit schematic in figure 3.13. It is a two-stage process; in the first stage, an analog integrator, whose output has been previously reset to zero, has the analog signal which is to be converted connected as its input signal. The input signal is integrated for a fixed time interval T_i and if it remains constant during this time the output of the integrator is a linear ramp.

The second stage of the process starts at the end of the time interval T_i at which instant the control logic disconnects the analog input signal and connects a reference voltage in its place as the integrator input. The reference voltage polarity is opposite to that of the analog input signal—it make the integrator output ramp back towards zero (a comparator is used to sense when it reaches zero). The time taken, T_r, for integration of the reference to bring the integrator output back to zero is directly proportional to the average value of the analog input signal during the interval T_i.

The time interval T_i is the time during which a fixed number of clock pulses are counted

$$T_i = N_i T_{cp}$$

where T_{cp} is the clock period and N_i is normally taken as the number of pulses which are required to fill and recycle the counter. The time interval T_r is measured in terms of the number of clock pulses counted in the time T_r

$$T_r = N_x T_{cp}$$

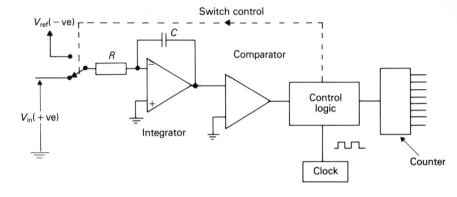

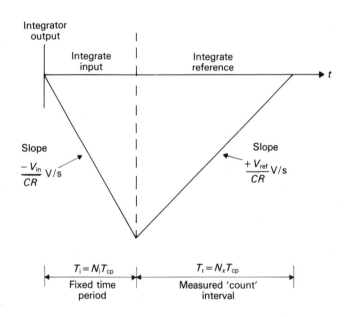

Figure 3.13 Simplified schematic for dual slope A/D converter

During the integrate signal phase the output of the integrator changes by an amount

$$V_o = \frac{1}{CR} \int_0^{T_i} V_{in} \, dt$$

where V_{in} is the analog input signal. During the integrate reference phase the output of the integrator is returned to zero. Thus

Analog to Digital Conversion Techniques

$$\frac{1}{CR}\int_0^{T_i} V_{in}\, dt - \frac{1}{CR}\int_0^{T_r} V_{ref}\, dt = 0$$

Substituting $T_i = N_i T_{cp}$, and $T_r = N_x T_{cp}$ and rearranging gives

$$N_x = \frac{N_i}{V_{ref}} \cdot \frac{\int_0^{N_i T_{cp}} V_{in}\, dt}{N_i T_{cp}}$$

or

$$N_x = \frac{N_i}{V_{ref}} \overline{V_{in}} \tag{3.12}$$

Where $\overline{V_{in}}$ is the average value of the analog input signal during the signal integrate phase of the conversion. The count N_x is recorded, it represents a digitally encoded form of the analog input signal.

The beauty of the dual-slope technique is that the theoretical accuracy depends only on the absolute value of the reference voltage and the equality of the individual clock periods during a conversion cycle. Only short term stability of clock frequency is required and this far easier to obtain than long term stability. Note that the component values, C and R, do not enter into the conversion relationship and it is therefore not necessary to use precise values. In the dual-slope conversion technique conversion time is not fixed but depends on the value of the analog input signal—this feature is a disadvantage in some types of application.

The dual-slope technique for A/D conversion has for a number of years been the preferred method for use in DPMs; the increasing size of the market for DPMs has encouraged continual developments in IC devices for use in dual-slope systems. Various refinements to the basic technique of figure 3.13 have been incorporated in the devices; these refinements include input buffering, auto-zeroing and auto-polarity. [7] Auto-zeroing involves an extra stage to the conversion cycle in which offsets in the analog circuitry are automatically balanced out. The auto-polarity function allows both positive and negative analog input signals to be converted; it involves the switching in of a reference voltage polarity opposite to that of the analog input signal.

IC devices which function as dual-ramp converter subsystems are available (for example, Motorola MC1505/MC14435, Intersil 8052/T101). Analog circuitry is contained in one device and digital circuitry in another. More recently single chip dual-slope converters have appeared on the market—they undoubtedly provide greater user convenience.

The ICL7106 (or 7107) is a device which contains all the active circuitry required for a $3\frac{1}{2}$ digit DPM (with auto-zero and auto-polarity). The device requires the addition simply of a display, four resistors, four capacitors and an input filter (if required) to make a working DPM. The basic circuit schematic of an LCD, DPM using the ICL7106 is shown in figure 3.14. The Intersil

66   **Data Converters**

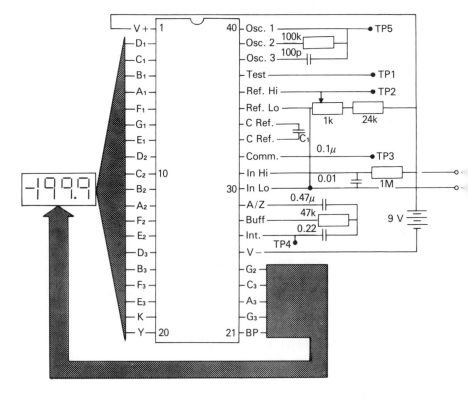

Figure 3.14   LCD digital panel meter using ICL 7106

DPM evaluation kit (ICL 7106 EV/KIT) contains the printed circuit board, converter chip, display and passive components; it gives a convenient way of practically looking into the dual-slope technique. At the same time it allows you to construct rapidly a working, high performance DPM; a photograph of the assembled kit is shown in figure 3.15. Test points are available at which signals

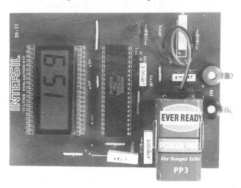

Figure 3.15   Photograph of DPM evaluation kit

Analog to Digital Conversion Techniques

can be monitored; figure 3.16 shows the waveform at the output of the analog integrator (pin 27) for two values of analog input signal. Note the fixed integrate signal time period and the constant slope but varying time period of the

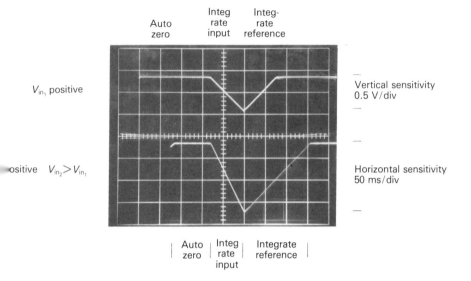

Figure 3.16 Integrator waveform (pin 27) in dual slope A/D converter for two values of V_{in}

integrate reference phase of the conversion. The measurement requires the use of a high impedance oscilloscope probe in order that loading should not interfere with the correct circuit operation.

3.10 QUANTISED-FEEDBACK A/D CONVERSION

The quantised-feedback and dual-slope methods for A/D conversion are somewhat similar in that both systems use a charge balancing technique. In the dual-slope system the charge supplied by the analog input signal during the fixed integrate signal time period is balanced by an equal and opposite charge supplied by the reference during the variable 'count' interval. The integrate signal and integrate reference and count occur as separate phases of the conversion process. However, in the quantised-feedback method, the integrate signal, integrate reference and count processes all occur simultaneously during a single fixed conversion time period.

Figure 3.17 is a simplified circuit schematic illustrating the functional operations underlying the quantised-feedback method. During conversion the sum of a continuous input current ($I_{in} = V_{in}/R_{in}$) and pulses of a reference current ($I_{ref} = V_{ref}/R_{ref}$) is integrated for a fixed number of clock periods. The reference current is of opposite polarity to the input current and is larger than the input

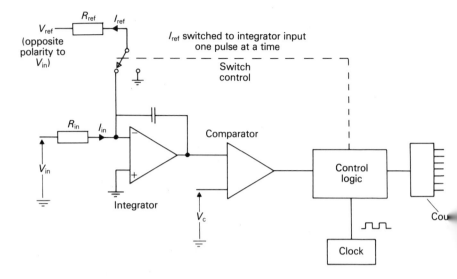

Figure 3.17 Simplified schematic for quantised feedback A/D converter

current ($I_{ref} \approx 2I_{in}$ full scale), it is switched in for exactly one clock period at a time, just frequently enough to maintain the integrator output near zero.

The comparator in the system senses when the output of the integrator exceeds a reference level (near zero) and signals the control logic to perform the clock switching of the reference current charge increments. The system maintains a continuous charge balancing between the charge supplied by I_{in} and that taken away by the I_{ref} pulses. The total number of I_{ref} pulses needed to maintain the balance during a fixed conversion time is counted and the count represents the digitally coded value of the analog input signal. Waveforms appearing at the integrator output in a practical quantised-feedback converter are shown in figures 3.19 and 3.20.

An expression for the conversion relationship in a quantised feedback converter is readily derived from the condition of zero net charge transfer. Let T_{cp} represent the clock period, then the magnitude of the I_{ref} charge increments are

$$q = \int_0^{T_{cp}} I_{ref} \, dt = \frac{V_{ref}}{R_{ref}} T_{cp} \tag{3.13}$$

the charge supplied by the continuous input current during the fixed conversion time period T_i is

$$Q = \int_0^{T_i} \frac{V_{in}}{R_{in}} \, dt \tag{3.14}$$

where $T_i = NT_{cp}$. (N is a fixed number). Equation 3.14 may be written as

Analog to Digital Conversion Techniques

$$Q = NT_{cp} \int_0^{NT_{cp}} \frac{V_{in}}{NT_{cp} R_{in}} \, dt$$

or

$$Q = \frac{NT_{cp}}{R_{in}} \overline{V_{in}} \qquad (3.15)$$

where $\overline{V_{in}}$ is the average value of the input signal during the conversion period. Q is balanced by N_x charge increments of I_{ref}. Thus

$$N_x q = Q$$

Substitution of values from equations 3.13 and 3.15 gives

$$N_x = \frac{R_{ref}}{R_{in} V_{ref}} N \overline{V_{in}} \qquad (3.16)$$

The count N_x represents a digitally encoded form of the average value of the input signal during the conversion period.

The quantised-feedback technique is insensitive to both long and short term drift in clock frequency since any change in clock frequency affects equally the charge supplied by the input signal and the charge supplied by the reference. The fixed conversion time of the technique is valuable if the converter is to be used in a data acquisition system—it allows the converter to be synchronised to the operation of the complete system.

Quantised-feedback A/D converters are available from several manufacturers: they come as two chip sets (for example Intersil LD111/110) with the analog and digital circuitry separated, and single chip devices which are now appearing on the market (for example Silcronix LD130). However, a design constraint with all single chip converters is the digital noise which inevitably crops up during a conversion and limits the maximum achievable sensitivity.

The Teledyne Semiconductor 8700 series of A/D converters are CMOS devices which use the quantised-feedback technique. They only require the addition of a few external passive components, power supplies and reference and provide a convenient method of practically examining a quantised-feedback converter at work. A test circuit is shown in figure 3.18. If you want to investigate the device fully and use it in a practical converter application you will need to obtain a copy of the device data sheet.

A simple test of the nature of the quantised-feedback mode of operation is to examine the waveform appearing at the output of the integrator (pin 15) of the device in figure 3.18. Typical integrator output waveforms for different values of the analog input signal are shown in figures 3.19 and 3.20. In figure 3.19a $I_{in} \ll I_{ref}$ and only 5 I_{ref} charge pulses are required during the conversion time in order to maintain charge balancing. Clock switching of the I_{ref} pulses is allowed when the integrator output exceeds approximately 1 V. In the second

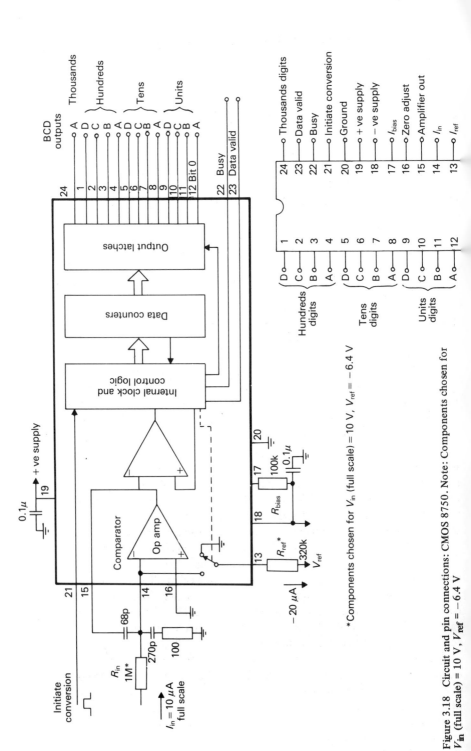

Figure 3.18 Circuit and pin connections: CMOS 8750. Note: Components chosen for V_{in} (full scale) = 10 V, V_{ref} = −6.4 V

Analog to Digital Conversion Techniques

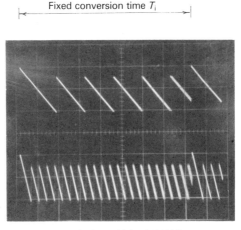

Vertical sensitivity 0.5 V/div;
horizontal sensitivity 0.2 ms/div

Figure 3.19 Integrator waveforms (pin 15) for quantised feedback A/D converter

trace figure 3.19b I_{in} is increased with a consequent increase in slope of the integrator ramp and a greater number of I_{ref} charge pulses are required to maintain the charge balance. In figure 3.20 I_{in} has been increased to a value approaching full scale ($I_{in} \cong I_{ref}/2$) and the time scale has been expanded so that the clock switching of the I_{ref} pulses can be seen clearly. Note that at times during the conversion the larger values of I_{in} cause the integrator output to appreciably exceed the comparator reference level before the next clocked pulse of I_{ref} can cause the return of the integrator output.

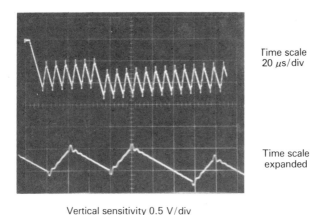

Time scale 20 μs/div

Time scale expanded

Vertical sensitivity 0.5 V/div

Figure 3.20 Integrator waveform (pin 15) for quantised feedback A/D converter $I_{in} = \frac{1}{2} I_{ref}$

3.11 PARALLEL A/D CONVERSION

The fastest A/D converters that are currently available make use of a parallel conversion technique (sometimes called flash encoding), in which all bit values are found simultaneously. A parallel converter uses a resistive potential divider to generate the series of voltage levels at which the transitions between adjacent codes in the conversion relationship occur. In an n-bit converter these voltage levels are $V_{ref}/2^{n+1}$ ($\frac{1}{2}$ LSB), $3V_{ref}/2^{n+1}$ ($1\frac{1}{2}$ LSB), $5V_{ref}/2^{n+1}$ ($2\frac{1}{2}$ LSB) ..., $(2^{n+1} - 3)V_{ref}/2^{n+1}$ $[(2^n - 2) + \frac{1}{2}$ LSB]. There are $2^n - 1$ levels for an n-bit conversion.

An outline schematic for an n-bit parallel converter is shown in figure 3.21.

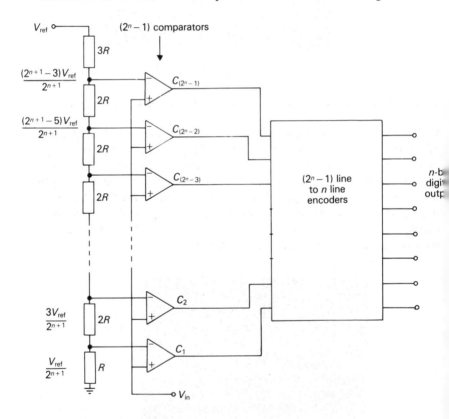

Figure 3.21 A/D conversion with a parallel conversion technique

The $(2^n - 1)$ voltage levels are used as the reference levels for $(2^n - 1)$ comparators, and the analog input which is to be converted is applied simultaneously to all comparators. If this analog input is increased from zero up to V_{ref} the comparator outputs go high in succession as the input exceeds their reference levels. Thus if $V_{in} < V_{ref}/2^{n+1}$ all comparator outputs are low. If

Analog to Digital Conversion Techniques

If

$$\frac{V_{ref}}{2^{n+1}} < V_{in} < \frac{3V_{ref}}{2^{n+1}}$$

$C_1 = 1$ and $C_{2,3,4,\ldots,(2^n-1)} = 0$

$$\frac{3V_{ref}}{2^{n+1}} < V_{in} < \frac{5V_{ref}}{2^{n+1}}$$

then $C_1 = C_2 = 1$ and $C_{3,4,5,\ldots(2^n-1)} = 0$, and so on.

Output logic circuitry is used to convert the $2^n - 1$ comparator states into an n-bit binary word. Since all bit values are found at the same time, the conversion time is limited only by the response time of the comparators and the propagation delay of the logic gates in the encoding system.

The parallel conversion principle is conceptually straight forward, but it has formidable practical difficulties associated with the large amount of circuitry which it requires to implement a high resolution conversion. Because of circuit complexity, the use of parallel converters has up until quite recently been limited to low resolution applications. Even a 4-bit conversion requires 15 matched high speed comparators and associated high speed logic circuitry. However, advances in LSI technology are overcoming the problems of circuit complexity and at the time of writing there are 8-bit parallel converters commercially available as single packet IC devices (for example TRW TDC1007J, an 8-bit parallel converter using ECL logic and capable of making 35×10^6 conversions/s. [8]

3.12 RATIOMETRIC A/D CONVERSIONS

A/D converters in effect compare an analog input signal with a reference voltage and generate a digital code word which represents the ratio V_{in}/V_{ref}. In A/D converter systems in which the reference voltage is externally applied or is externally available, the ratiometric nature of an A/D conversion can sometimes be used to advantage.

Many measurement transducers are essentially ratiometric in that it is the ratio of their output signal to a reference signal that is proportional to the physical parameter being measured. If the output of such a transducer is to be converted into digital form, the digital output is made independent of reference voltage variations if the converter and transducer share the same reference voltage supply.

A potentiometer used as a position sensor is an easily understood example of a ratiometric transducer and is shown in figure 3.22. The ratio of the potentiometer output voltage to the potentiometer supply voltage is equal to the distance of the wiper from one end expressed as a fraction of the full wiper traverse. In figure 3.22 the analog output voltage produced by the potentio-

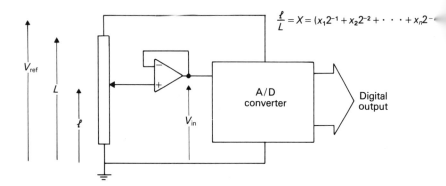

Figure 3.22 Digital output independent of reference voltage variations when ratiometer transducer and converter share same reference

meter is converted into digital form. Assuming a natural binary conversion code we may write

$$\left(\frac{V_{in}}{V_{ref}} + \frac{1}{2^{n+1}}\right) > (x_1 2^{-1} + x_2 2^{-2} + \ldots + x_n 2^{-n}) > \left(\frac{V_{in}}{V_{ref}} - \frac{1}{2^{n+1}}\right)$$

However, the converter and potentiometer share the same reference voltage and the potentiometer ratio $X = l/L = V_{in}/V_{ref}$ Thus

$$\left(X + \frac{1}{2^{n+1}}\right) > (x_1 2^{-1} + x_2 2^{-2} + \ldots + x_n 2^{-n}) > \left(X - \frac{1}{2^{n+1}}\right)$$

The binary digital output code gives the fractional value of X with a quantisation uncertainty of $\pm \frac{1}{2}$ LSB. The output is independent of reference voltage variations.

Resistive transducers are used in the measurement of a wide range of physical parameters: temperature, pressure, light intensity, etc. The resistive transducer is normally made one arm of a balanced bridge. Changes in the physical variable to which the transducer is sensitive cause a bridge unbalance, the extent of the unbalance being used to measure the change in the physical variable. Bridge unbalance output depends on bridge supply voltage. If the bridge output is converted into a digital signal, the digital signal can be made independent of bridge supply by a ratiometric conversion in which the bridge supply acts as the converter's reference signal.

The above principles are illustrated in figure 3.23. The bridge produces a differential output signal. A differential input measurement amplifier is shown being used to convert this differential signal into a single ended signal before

Analog to Digital Conversion Techniques

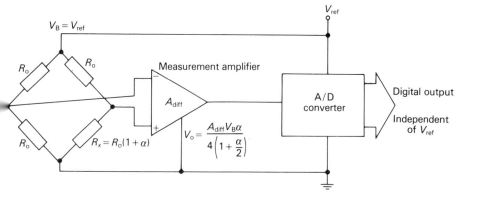

Figure 3.23 Ratiometric conversion removes bridge supply dependence in resistive bridge measurements

applying it as the input signal to an A/D converter. Differential input measurement amplifiers provide a high degree of rejection of unwanted common mode interference signals. [1,9]

Operational amplifiers can often be used to condition resistive transducer signals as an alternative to a fully developed measurement amplifier in applications less demanding of high common mode rejection. [1] Figure 3.24 shows the successive approximation converter system previously given in figure 3.10,

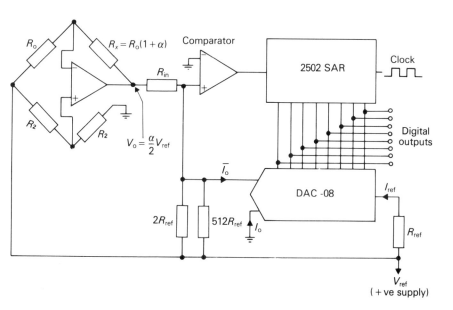

Figure 3.24 Ratiometric conversion gives digital read-out of register deviation

combined with such an operational amplifier configuration. The system provides a digital output which depends on the fractional deviation of a resistance value from a standard value.

The operational amplifier in figure 3.24 produces an output signal $V_o = -(\alpha/2) V_{ref}$ [1] – a negative voltage if $R_x > R_o$ and a positive voltage if $R_x < R_o$. The comparator, DAC–08 and 2502 successive approximation register form a successive approximation converter system with a complementary offset binary code. Digital code values are related to mid-quantisation range of values by

$$-\frac{\alpha}{2} V_{ref} = \frac{V_{ref}}{R_{ref}} R_{in} [(x_1 2^{-1} + x_2 2^{-2} + \ldots + x_8 2^{-8}) - \tfrac{1}{2}]$$

The operational amplifier and converter share the same reference and with $2R_{in} = R_{ref}$ this gives

$$-\alpha = (\bar{x}_1 2^{-1} + \bar{x}_2 2^{-2} + \ldots + \bar{x}_8 2^{-8}) - \tfrac{1}{2} \tag{3.17}$$

Some code values are as follows. (An offset binary code which gives the sign and fractional value of α, the resistor deviation, with quantisation uncertainty ± 1/512.)

α as Analogue Fraction	Bit Values							
	x_1	x_2	x_3	x_4	x_5	x_6	x_7	x_8
$+\frac{128}{256} = \frac{1}{2}$	1	1	1	1	1	1	1	1
$+\frac{127}{256}$	1	1	1	1	1	1	1	0
.								
.								
$+\frac{1}{256}$	1	0	0	0	0	0	0	0
0	0	1	1	1	1	1	1	1
$-\frac{1}{256}$	0	1	1	1	1	1	1	0
.								
.								
$-\frac{126}{256}$	0	0	0	0	0	0	0	1
$-\frac{127}{256}$	0	0	0	0	0	0	0	0

The normal and complementary output currents of a DAC–08 can be used as a means of nulling resistive bridges. The system given in figure 3.25 shows how this can be accomplished. [10] The system provides a digital representation of the bridge's out of balance current–it does this without using an operational amplifier.

The DAC–08 up/down counter and comparator form a tracking A/D con-

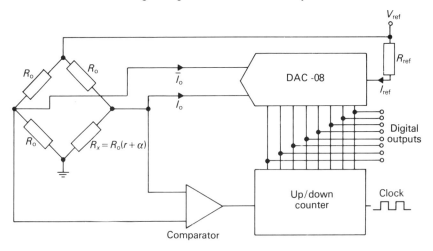

Figure 3.25 Tracking converter nulls bridge and gives digital read-out of bridge out-of-balance current output

verter system. The DAC reference and the bridge are supplied by the same voltage and the system works in the ratiometric mode. The comparator controls the counting mode (up or down) in such a way that the system seeks to bring the comparator input terminals to the same potential and nulls the bridge out of balance voltage.

The relationship between digital output and resistance deviation α is approximately linear only for small values of α. The relationship which gives nominal code values is

$$\alpha = 2\,\frac{R_o}{R_{ref}}\left[2(x_1 2^{-1} + x_2 2^{-2} + \ldots + x_8 2^{-8}) - \frac{255}{256}\right] \qquad (3.18)$$

for $\alpha \ll 1$. Equation 3.18 represents a symmetrical offset binary code. It should be stressed that equation 3.18 is valid only when resistor values are such that the bridge is close to balance. The relationship between digital output and resistor deviation α becomes markedly non-linear when the bridge is well off balance. Also, like all bridge current read-out systems there is a dependence of output on bridge impedance level, R_o.

3.13 COMPARISON OF A/D CONVERSION TECHNIQUES

A brief summary is now given in which the main characteristics of devices which use the various conversion techniques that have been discussed will be compared. There are normally three dominant factors which govern the choice of the A/D converter to be used in an application: speed, accuracy and cost. It is convenient to compare the different techniques in relation to these three factors.

Speed limitations are inherent in the particular techniques but a range of speeds can be expected for different converter devices using the same conversion technique. A speed/accuracy design compromise is inherent in all the conversion techniques. Increased accuracy is obtained at the expense of a reduction in conversion speed. In general, cost is directly related to speed and accuracy, but the cost of a converter device (indeed of any device), is greatly influenced by market factors which might be quite unrelated to the performance of the device. The bigger the market is, the more firms there will be attempting to get a slice of it, and competition brings down the cost.

As far as inherent speed limitations are concerned the techniques discussed in this chapter in order of decreasing speed (increase in conversion time) are: parallel conversion, successive approximation conversion, tracking conversion and integrating conversion.

If sheer conversion speed is a dominating factor in an application, the designer will have to choose a converter which uses the parallel conversion technique. At the time of writing there are not many ultra-fast converters on the market and they tend to be expensive, particularly the higher resolution devices. Some firms which make very fast converters are TRW LSI Products, Datel-Intersil, Computer Labs and Motorola. Available parallel converters achieve an 8-bit conversion time of the order of 30 ns. Applications in which conversion time must be very short are increasing for example, radar, digital TV and fast transient event recorders. [11, 12] No doubt this growing range of applications will interest more manufacturers in very high speed converters and will result in a greater number of less costly devices appearing on the market.

In converters which use the successive approximation technique conversion time is related to the number of bits (to the resolution). The fastest successive approximation converters currently available are 8-bit devices which have a conversion time of the order of 0.8 μs. General purpose 8-bit devices may be expected to have conversion time of the order of 30 μs. High resolution 16-bit successive approximation converter modules have conversion times of the order of 400 μs.

Successive approximation is probably the most widely used technique in the A/D converters which are currently available; there are a large number of manufacturers making many different successive approximation devices. The successive approximation technique has more possible sources of error than the slower integrating techniques. The highest performance successive approximation converters are usually the discrete component modular type. Costs rise steeply if the application calls for state-of-the-art speed and accuracy. Successive approximation converters must be preceded by a sample/hold module when they are used to digitise rapidly changing analog signals (see section 5.3).

Tracking converters, unlike successive approximation types, do not require an input sample/hold. Although they use simpler digital logic circuitry than successive approximation types they are subject to similar errors, but do not have the speed of a successive approximation converter preceded by a sample/

Analog to Digital Conversion Techniques 79

hold (see section 5.8.2). There are comparatively few tracking or counting-type converters on the market compared with successive approximation and integrating types.

Integrating converters are the slowest type but they are capable of very high accuracy. Conversion times of currently available integrating types range from 0.3 ms for a fast 8-bit converter to 250 ms for a slow 16-bit one. The integrating technique has the advantage of averaging-out noise components of the analog input signal whose frequencies are higher than the inverse of the integration period. Integrating converters do not require sample/holds. If the application is one in which the analog signals are slowly varying an integrating converter will provide the greatest accuracy for the least cost. Integrating converters are available from a large number of manufacturers.

3.14 SELF-ASSESSMENT EXERCISES

1. From the list of functional units given below pick out those which are essential elements of
   (a) a tracking A/D converter
   (b) a successive approximation A/D converter
   (c) a parallel A/D converter
   (d) a dual slope A/D converter
comparator; integrator; clock sources; logic encoder; successive approximation register; digital counter; DAC; digital shift register.

2. Of the A/D conversion techniques described in this chapter, name those techniques in which conversion time is independent of analog signal magnitude.

3. Which of the following statements are true?
   (a) The conversion time of a successive approximation converter depends on the number of bits in the conversion code.
   (b) The conversion time of a parallel converter depends on the number of bits in the conversion code.
   (c) In a dual-slope A/D converter: (i) the input signal is integrated for a fixed time interval; (ii) the reference signal is integrated for a fixed time interval; (iii) the conversion takes a fixed time.
   (d) Conversion accuracy in a dual-slope converter depends on: (i) short term clock frequency stability; (ii) accuracy of integrator components; (iii) reference voltage stability.
   (e) The digital output of a successive approximation A/D converter is insensitive to high frequency noise components of the analog input signal.
   (f) The digital output of a dual-slope A/D converter is insensitive to high frequency noise components of the analog input.

(g) Valid digital data is always present at the output of a successive approximation A/D converter

4. How many comparators are required in a 6-bit parallel A/D converter?

5. Fill in the blanks in the following statements.
   (a) A successive approximation register is a special purpose .... data converter.
   (b) An output digital storage register is added to a successive approximation converter in order to ....
   (c) The preferred conversion technique for use in digital panel meters is ....
   (d) The fastest A/D conversion technique is ....
   (e) The digital output of a dual-slope converter is obtained by counting the number of clock pulses which occur while the .... signal is integrated.

6. The following components are used in figure 3.2, R_{ref} = 10 kΩ, R_{in} = 5 kΩ, V_{ref} = 10 V, clock frequency = 2 MHz. Find the conversion time and the digital code for
   (a) a nominal full scale analog input signal
   (b) an input signal 4.5 V.

7. A 6-bit current output DAC is used in figures 3.3b and 3.3c; the output current determined by $I_o = I_{ref}(x_1 2^{-1} + x_2 2^{-2} + \ldots + x_6 2^{-6})$; $I_{ref} = V_{ref}/R_{ref}$ is to be set at 2 mA. If V_{ref} = 10 V find the resistor values required for a ± 10 V full scale analog input range. If a clock signal frequency 1 MHz is used, find the loop slew rate and the maximum frequency for a ± 5 V peak-to-peak sinusoidal input signal so as not to exceed the loop slew rate. Which circuit has the highest input impedance?

8. Give the trial guesses which you would make in order to find the value of an unknown number between 0 and 64 with no more than six guesses. After each guess you are told whether your guess is high or low. Illustrate your answer by taking the unknown numbers as 11 and 53 and any other number of your choice. Compare the above guessing sequence to the functional operation performed by a successive approximation A/D converter.

9. An analog input signal of 4.5 V is applied to the successive approximation converter system given in figure 3.8. Draw up a table (similar to that given in section 3.5) which shows the register bit values, values of the DAC output current as a fraction of I_{ref} and the comparator output states produced in a conversion sequence. Sketch the 2502 timing diagram (similar to figure 3.7) for the conversion sequence.

Analog to Digital Conversion Techniques 81

10. In the successive approximation converter system given in figure 3.10 $V_{ref} = 10$ V, $R_{ref} = 5$ kΩ and $R_{in} = 10$ kΩ. The DAC-08 $\overline{I}_o$ output current is offset by a resistor value $2R_{ref}\|512R_{ref}$ connected between V_{ref} and the DAC-08 pin 2. This gives the system a bipolar capability and centres the $\pm \frac{1}{2}$ LSB quantisation uncertainty. An input signal -3 V is applied to the system; indicate the register bit values, values of $\overline{I}_o/I_{ref}$ and comparator output states produced in a conversion sequence, and draw up a table similar to that given in section 3.5. Give the code values for full scale positive, zero and full scale negative and give the analog input signals which produce these code values.

11. A 10-bit natural binary coded DAC, a successive approximation register and a comparator are connected to form a successive approximation converter. What is the conversion time of the system if it is supplied by a 2 MHz clock?

12. A $3\frac{1}{2}$-digit decimal display dual-slope converter is based on the schematic given in figure 3.13. The system uses a counter formed by a cascade of three BCD quads plus a single flip flop. The input signal is integrated for the time interval required to count 1000 clock pulses (for the time taken by the three BCD quad cascade to fill and recycle).

What value of reference voltage must be used if the full scale count recorded during the integrate reference phase of the conversion (1999) is to represent a full scale analog input signal 1.999 V?

What is the highest clock frequency which will allow the system to achieve maximum rejection of: (a) 50 Hz, (b) 60 Hz, pick up?

If the system uses an integrator capacitor $C = 0.1$ µF and an integrator resistor $R = 100$ kΩ sketch the integrator output voltage variation which you expect to take place when: (a) a full 1.999 V, (b) a 0.500 V input signal is applied? Assume the reference voltage and clock frequency calculated in the first part of the question are used.

Suggest system parameter changes for a 19.99 V full scale input signal. Note: integrator CR value must be changed to avoid excessive integrator output voltage swing.

13. A DAC-08, a 2502 SAR, and a comparator are to be used to implement a 6-bit successive approximation A/D conversion. The system is to have a full scale analog range ± 10 V and a complementary offset binary code. Sketch a suitable interconnector layout giving component values and introducing any other functional elements that may be required in the system.

14. The system shown in figure 3.24 is to be used to provide a digital read-out of the out-of-balance signal produced by a resistive bridge. If $R_{in} = R_{ref}/2$ and $R_2 = R_o = 10$ kΩ, what digital output is produced when R_x is (a) 12 kΩ, (b) 9 kΩ?

4 Digital to Analog Converter Applications

Until quite recently there has tended to be a clear distinction between analog and digital electronic systems. Functional operations such as amplification, waveform generation, filtering, modulation and demodulation, etc., have traditionally been regarded as purely analog operations; In many cases the implementation of such functions has been founded on operational amplifier based circuitry. [1] Logical operations involved in process control and mathematical operations performed on abstract digital numbers have been the main area of application for digital electronic systems. Digital systems are founded on circuitry using gates, flip flops, counters, shift registers, etc.

One of the main reasons for the divisions of electronic systems between analog and digital has been the high cost of data converters. Data converters are the essential interface components which allow analog and digital devices to work together. The inexpensive monolithic data converters which are now available make it economically viable to combine analog and digital devices. Such a combination opens up many versatile new techniques of signal manipulation to the system designer with a knowledge of both analog and digital devices. Data converters permit a precise digital control to be added to analog functional applications; they also allow many of the traditionally analog functional operations to be performed with circuitry based on digital devices. This chapter gives examples of some of the many analog/digital applications that are possible with the use of data converters.

4.1 DIGITALLY PROGRAMMED VOLTAGE AND CURRENT SOURCES

In electronic instrumentation systems there is often the need for a well-regulated voltage and/or current source. Operational amplifier feedback circuits can be used to provide such sources. [1] A DAC can be combined with an operational amplifier circuit to add the convenience of a precise digital setting of the source. Digital setting can be done manually (say with a BCD thumbwheel switch) or completely automatically under microprocessor control. Microprocessor controlled voltage and current sources are useful in automatic testing and control systems.

Digital to Analog Converter Applications

A voltage output DAC is in itself a digitally controlled voltage source. Output current is proportional to the product of the DAC reference voltage and the binary word which is applied to the DAC's digital inputs. Scaling can be set by choice of the feedback resistor which is connected to the DAC's output operational amplifier. It is a simple matter to increase the amplifier's output capabilities by including some form of booster within the feedback loop of the amplifier. [1] A bipolar voltage source with offset binary coding of the digital control input can be arranged by offsetting the DAC's output by half full scale. The above principles are illustrated by the functional schematic given in figure 4.1. The magnitude of the voltage source is determined by the equation

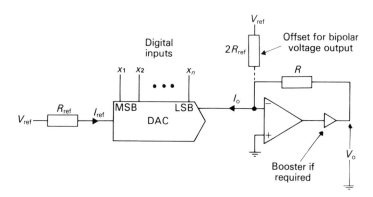

Figure 4.1 Digitally programmed voltage source

$$V_o = \frac{V_{ref}}{R_{ref}} R \left(x_1 2^{-1} + x_2 2^{-2} + x_3 2^{-3} + \ldots + x_n 2^{-n} \right) \quad (4.1)$$

where the bit values x_i are either '0' or '1'. With the DAC's output offset by half scale the source has a value

$$V_o = \frac{V_{ref}}{R_{ref}} R \left[\left(x_1 2^{-1} + x_2 2^{-2} + \ldots + x_n 2^{-n} \right) - \tfrac{1}{2} \right] \quad (4.2)$$

In applications in which the digital inputs are manually set, denary numbers are generally preferred to binary numbers. Manual setting of digital inputs can be performed with a BCD thumbwheel switch module and a BCD DAC (for example, Precision Monolithics DAC −20) is required in the circuit.

A current output DAC functions as a digitally programmed current source. Design considerations in practical programmed current source applications centre on configuring an output operational amplifier so that it provides a current output and at the same time gives current gain. A variety of operational amplifier current sources have been devised. [1] The type of circuitry used is generally governed by load requirements. An important consideration is whether or not the load must be earthed or can be allowed to float.

The circuit shown in figure 4.2 is a simple programmable current source suitable for loads that are not connected to earth. The performance equation for the circuit is readily derived if we make use of the normal ideal amplifier assump-

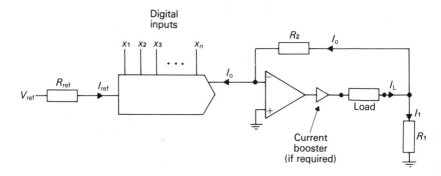

Figure 4.2 Digitally programmed current source (floating load)

tions. The amplifier maintains a virtual earth at its inverting input terminal and forces the DAC's output current to flow through resistor R_2. The following relationships hold

$$I_o R_2 = I_1 R_1$$

and

$$I_L = I_1 + I_o = I_o \left(1 + \frac{R_2}{R_1}\right)$$

I_o is the DAC output current controlled by the values of the digital inputs which are applied to it

$$I_o = \frac{V_{ref}}{R_{ref}} \left(x_1 2^{-1} + x_2 2^{-2} + \ldots + x_n 2^{-n}\right)$$

Thus

$$I_L = \frac{V_{ref}}{R_{ref}} \left(1 + \frac{R_2}{R_1}\right) \left(x_1 2^{-1} + x_2 2^{-2} + \ldots + x_n 2^{-n}\right) \qquad (4.3)$$

As before, if the current output capabilities of the operational amplifier are inadequate an inside the loop current booster can be used. The range of load impedances that can be driven are governed by the amplifiers output voltage limits in accordance with the relationship

$$I_L Z_L + I_1 R_1 \leqslant V_{o\,max}$$

Digital to Analog Converter Applications 85

A circuit for a programmable current source suitable for driving a grounded load is given in manufacturers' application notes. [13] A much simpler circuit configuration is shown in figure 4.3. DAC's like Precision Monolithics DAC–08 give a current output at high output impedance and with wide voltage compliance. The high output impedance allows the use of a modified operational amplifier current inverter to provide both current gain and a current source output. [1] In figure 4.3 a simple emitter follower is used to boost the operational amplifier's output current. The amplifier, in maintaining its input terminals at the same potential, forces the relationship

$$I_L R_1 = I_o R_2$$

The load current $I_L = (I_o R_2 / R_1)$ and current gain is thus provided. As before the DAC's output current I_o is fixed by the TTL levels applied to its digital inputs

$$I_o = I_{ref} \left(x_1 2^{-1} + x_2 2^{-2} + \ldots + x_8 2^{-8} \right)$$

Using the component values shown in figure 4.3 the load current is 10 times the DAC's output current. With the DAC's reference current set at 2 mA load current is digitally programmable in the range 0 to 20 mA.

The design is tolerant of minor power supply variations. Output voltage compliance is approximately the operational amplifier's output voltage limit less the voltage drop across R_1. ($V_{o_{max}}^+ - I_L R_1$). Assuming $V_{o_{max}}^+$ is say 12 V the maximum load resistor value is 500 Ω. Output settling time is largely controlled by the amplifier's slew rate. Full scale settling into 500 Ω takes approximately 2 μs.

The circuit can be modified to give a bipolar output current with a symmetrical offset binary code for the digital input levels. The modification which is shown in figure 4.4 makes the use of the DAC–08's $\overline{I_o}$ output current.

In Figure 4.4 the operational amplifier forces the relationship

$$I_1 R_1 = I_o R_2 - \overline{I_o} R_3$$

The current $I_1 = \overline{I_o} + I_L$; the load current is thus

$$I_L = I_o \frac{R_2}{R_1} - \overline{I_o} \frac{R_3 + R_1}{R_1}$$

If resistor values are chosen so that $R_3 = R_2 - R_1$ this makes

$$I_L = \left(I_o - \overline{I_o} \right) \frac{R_2}{R_1}$$

A simple complementary emitter follower (transistors Q_1, Q_2) is used to boost the operational amplifier's output current.

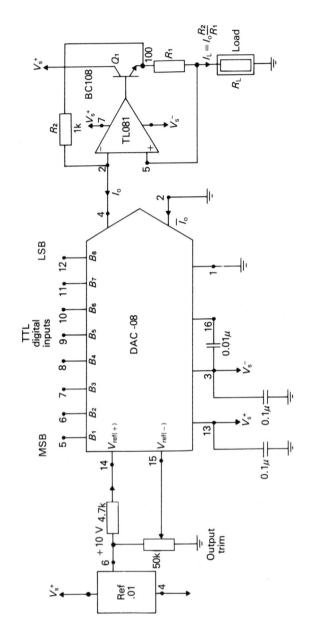

Figure 4.3 Programmable current source (unipolar)

Digital to Analog Converter Applications

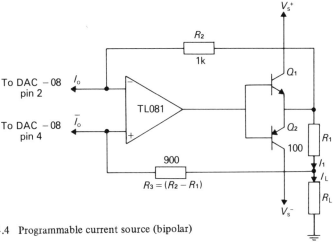

Figure 4.4 Programmable current source (bipolar)

4.2 DIGITAL GAIN CONTROL

Multiplying DACs can be used to provide a digital control of analog signal gain (or attenuation). Control of gain for unipolar analog signals is straightforward — the analogue signal is simply made to act as the reference voltage for the multiplying DAC. The technique is illustrated by the digitally programmed voltage source circuit shown in Figure 4.1. With $V_{ref} = V_{in}$ the circuit performance equation is

$$\frac{V_o}{V_{in}} = \frac{R}{R_{ref}} \left(x_1 2^{-1} + x_2 2^{-2} + \ldots + x_n 2^{-n} \right) \quad (4.4)$$

Gain is controlled by the digital input word applied to the DAC.

The processing of a bipolar analog signal with a multiplying DAC requires rather more circuitry. In most multiplying current output DACs the input reference current and analog output current are unidirectional. A bipolar analog signal can be accommodated as a variable reference only if it is used as a means of modulating a steady quiescent reference current. The modulated reference gives a corresponding modulation to the DAC's analog output current. The amplitude of the output modulation and the quiescent output current level about which it takes place are both controlled by the digital input word applied to the DAC.

A.C. coupling the DAC's output (after it has been converted to a voltage) with a d.c. blocking capacitor can be used to remove the quiescent output level. This method gives a transient output disturbance when a gain change by the digital input causes a change in the quiescent output. An alternative technique is needed in applications requiring a response down to zero frequency. A second DAC is used to subtract out completely the quiescent component of the output signal.

Data Converters

A direct coupled digital gain control circuit is illustrated in figure 4.5. [10] The reference current of the upper DAC (DAC 1) is modulated by the analog input signal

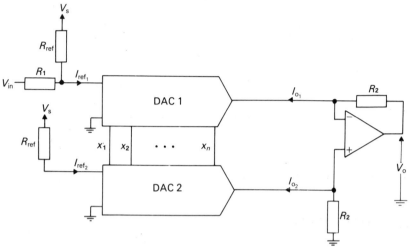

Figure 4.5 Direct coupled digital gain control (digital attenuator)

$$I_{ref_1} = \frac{V_s}{R_{ref}} + \frac{V_{in}}{R_1}$$

where V_{in} can take both positive and negative values ($V_{in} < (V_s R_1)/R_{ref}$). DAC 2, the lower DAC, has its reference current fixed at a value

$$I_{ref_2} = \frac{V_s}{R_{ref}}$$

Both DACs have the same digital word applied to their digital inputs and produce output currents

$$I_{o_1} = I_{ref_1} \left(x_1 2^{-1} + x_2 2^{-2} + \ldots + x_n 2^{-n} \right)$$

and

$$I_{o_2} = I_{ref_2} \left(x_1 2^{-1} + x_2 2^{-2} + \ldots + x_n 2^{-n} \right)$$

The operational amplifier acting as a differential input current to voltage converter, [1] produces an output voltage

$$V_o = (I_{o_1} - I_{o_2}) R_2$$

Substitution for I_{o_1} and I_{o_2} gives rise to the expression for the output voltage as

Digital to Analog Converter Applications

$$V_o = V_{in} \frac{R_2}{R_1} \left(x_1 2^{-1} + x_2 2^{-2} + \ldots + x_n 2^{-n} \right) \quad (4.5)$$

Output polarity is controlled by the analog input polarity. Gain magnitude is controlled by the natural binary coded digital word applied to the DAC's digital inputs. The circuit, in effect, performs a two-quadrant multiplication. It multiplies a bipolar analog signal with a unipolar digital signal.

Note that it is possible to implement a logarithmic digital gain control by using logarithmic DACs in place of linear DACs (for example, Precision Monolithics DAC–76).

There are multiplying DACs available which are designed for use with bipolar reference voltages and in which the polarity of the output bit currents are controlled by the reference voltage polarity. Analog Devices AD7500 series of multiplying DACs are such devices; they are well suited for use in programmed gain and attenuation applications. [14] Programmed gain circuits which can accommodate bipolar analog signals require the use of only one of these devices.

A functional schematic for the AD7520 is given in figure 4.6. The external connections required to multiply a bipolar analog signal (± 10 V) by a natural binary coded fraction are shown. The circuit functions as a digitally programmed attenuator. The AD7520 consists essentially of a 10-bit length R-$2R$ resistor ladder network and 10 CMOS current switches. The switches steer the bit currents to either I_{o_2} line (ground) or the I_{o_1} output line. The switches are controlled by the logic levels which are applied to the digital inputs. An externally connected operational amplifier is configured as a current to voltage converter. It maintains a virtual ground at the I_{o_1} output pin and converts the I_{o_1} current to an output voltage. A resistor R formed on the AD7520 chip serves as the operational amplifier's feedback resistor.

The analog input signal serves as the AD7520 reference voltage. With the I_{o_1} output pin maintained at ground potential the output current is fixed by the ladder network and switch positions at a value

$$I_{o_1} = \frac{V_{in}}{R} \left(x_1 2^{-1} + x_2 2^{-2} + \ldots + x_{10} 2^{-10} \right) \quad (4.6)$$

The operational amplifier converts this current to an output voltage $V_o = -I_{o_1} R$. Substitution gives the circuit performance equation as

$$V_o = -V_{in} \left(x_1 2^{-1} + x_2 2^{-2} + \ldots + x_{10} 2^{-10} \right) \quad (4.7)$$

The dynamic response characteristics (bandwidth, slew rate and offset) are determined by the type of operational amplifier used. [1] If an externally frequency compensated amplifier is used the recommended unity gain frequency compensation should be applied to it.

In the circuit configuration of figure 4.6 the DAC (the R-$2R$ network) acts as the operational amplifier's input resistor. A slight modification puts the DAC

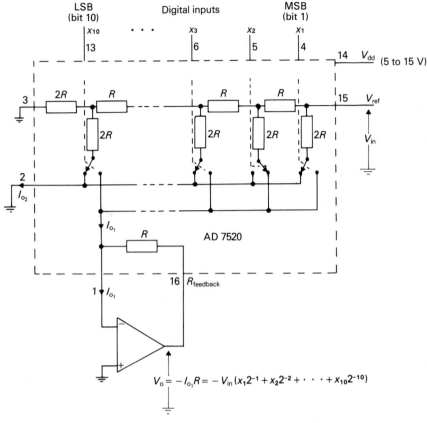

Figure 4.6 Digitally programmed attenuation (courtesy of Analog Devices)

in the feedback path between the output and the inverting input of the amplifier. The modified arrangement is shown in figure 4.7. It produces an output voltage which is equal to the analog input signal divided by the value of the natural binary coded digital number (binary fraction) applied to the DAC's digital inputs.

The output voltage of the operational amplifier in figure 4.7 acts as the DAC reference voltage. The DAC output current I_{o_1} is fed back to the inverting input terminal of the operational amplifier and the resistor formed on the AD7520 chip, R_{FEEDBACK}, serves as the operational amplifier's input resistor. The operational amplifier in maintaining a virtual ground at its inverting input forces equality between the magnitude of the input current, V_{in}/R, and the DAC output current I_{o_1}. Now

$$I_{o_1} = \frac{V_o}{R}\left(x_1 2^{-1} + x_2 2^{-2} + \ldots + x_{10} 2^{-10}\right)$$

The operational amplifier forces

Digital to Analog Converter Applications

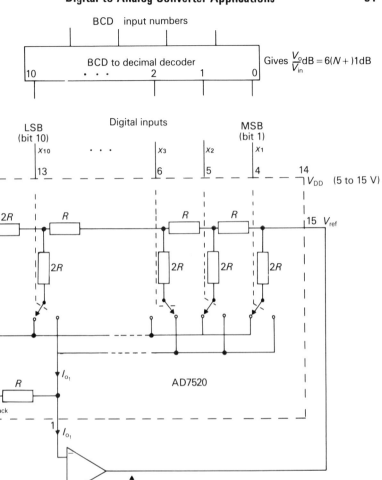

Figure 4.7 Digitally programmed gain

$$\frac{V_{in}}{R} = -I_{o1}$$

Substitution and rearrangement gives the circuit performance equation as

$$V_o = -\frac{V_{in}}{(x_1 2^{-1} + x_2 2^{-2} + \ldots + x_{10} 2^{-10})} \quad (4.8)$$

With the bit switches in the position shown in figure 4.7, ($x_3 = 1, x_{124...10} = 0$) the circuit has a gain of -8.

The addition of a BCD to decimal decoder provides a programmed gain circuit with 10 different gain settings. [15] A 4-bit BCD input applied to the decoder causes only one of the AD7520 digital inputs to go high. The closed loop signal gain of the operational amplifier can then be expressed as -2^{N+1}, where $N = 0, 1, \ldots, 9$, designates the decoder output which is high. Expressed in dB the signal gain is

$$G \simeq 6(N+1) \text{ dB}$$

(Note that $20 \log_{10} 2 = 6.02$.) (4.9)

As in other DAC circuits in which an operational amplifier converts an output current to an output voltage, it is the operational amplifier in figure 4.7 which largely determines the dynamic behaviour of the system. In operational amplifier feedback configurations, switching of the closed loop signal gain between different values is normally brought about by switching the value of the feedback fraction in the circuit. Change of feedback fraction causes a change in loop gain and a change in closed loop signal bandwidth. In the programmed gain circuit of figure 4.7 the feedback fraction, loop gain and signal bandwidth decrease as the gain is switched to higher values.

The accuracy of operational amplifier based circuitry is influenced by the value of the amplifier's d.c. offsets. In the circuits shown in figure 4.6 and 4.7, offset balancing, if used, should be performed using the internal null scheme recommended for the particular amplifier type used.

Note: Readers who are unfamiliar with the factors influencing the accuracy of operational amplifier circuits would be well advised to consult a specialised text on operational amplifiers. [1]

4.3 ARITHMETIC OPERATIONS WITH DIGITAL AND ANALOG VARIABLES

Low cost monolithic multiplying DACs can be used to advantage to perform various arithmetic operations such as addition, subtraction, multiplication and division involving analog and digital variables. In the digital gain control circuits described previously (figures 4.5 and 4.6) a bipolar analog signal is multiplied by a unipolar digital signal to give a bipolar analog output; this is a two-quadrant multiplication. Two-quadrant multiplication with unipolar analog and bipolar digital inputs and also full four-quadrant multiplier circuits can be arranged. Sometimes the requirement is to perform arithmetic operations with digital variables and to present the result of such operations in analog form. Digital ICs can be used to perform digital arithmetic operations and a DAC used to convert the result. Alternatively, DACs can themselves be configured to act as digital arithmetic building blocks, an approach that can often reduce the number of IC

Digital to Analog Converter Applications

packages required in an application. Circuits which illustrate some of the arithmetic operations that can be performed with DACs are now given. [10, 16]

4.3.1 Analog Variable Multiplied by Digital Variable with Result in Analog Form

The basic operation performed by most multiplying DACs is a single quadrant multiplication: a unipolar analog signal (the DAC's reference voltage) is multiplied by a unipolar digital signal to give a unipolar analog output signal. A multiplying DAC normally requires additional circuitry if it is to be made to perform two and four-quadrant multiplication.

Two-quadrant multiplication of a unipolar analog signal by an offset binary coded bipolar digital signal is readily achieved: the DAC's analog output is simply offset by an amount equal to half full scale. The technique is illustrated by the digitally programmed voltage source given in figure 4.1: the fixed reference voltage in figure 4.1 is replaced by the unipolar analog input signal. The use of a DAC device that gives complementary current outputs can give a symmetrical offset binary code to the bipolar digital signal. A circuit arrangement is shown in figure 4.8, where the bipolar analog output signal is provided by an operational amplifier, configured as a differential input current to voltage converter.

The output voltage produced by the circuit in figure 4.8 is

$$V_o = (I_o - \overline{I_o}) R_2$$

where

$$I_o = \frac{V_{in}}{R_1} \left(x_1 2^{-1} + x_2 2^{-2} + \ldots + x_8 2^{-8} \right)$$

and

$$\overline{I_o} = \frac{255}{256} \frac{V_{in}}{R_1} - I_o$$

Substitution gives the expression for the output voltage

$$V_o = V_{in} \frac{R_2}{R_1} \left[2 \left(x_1 2^{-1} + x_2 2^{-2} + \ldots + x_8 2^{-8} \right) - \frac{255}{256} \right] \quad (4.11)$$

This represents a symmetrical offset binary code; some values are tabulated in Figure 4.8.

The system shown in figure 4.9 gives a four-quadrant multiplication. It multiplies a bipolar analog signal by a bipolar digital signal (symmetrical offset binary code) to give a bipolar analog output. Output polarity reflects the polarity of both inputs. The output is in the form of a differential current $(I_1 - I_2)$; the operational amplifier shown in figure 4.9 converts this differential current to a single ended output voltage. Balanced loads (for example, a transformer or bridge) can be driven directly by the differential output current without the

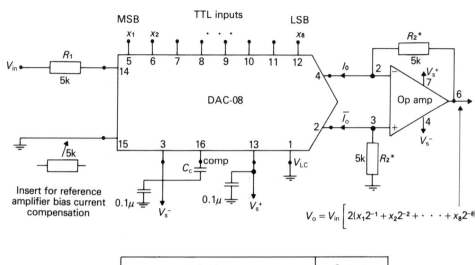

	Digital inputs								Output
	X1	X2	X2	X4	X5	X6	X7	X8	
Full scale positive	1	1	1	1	1	1	1	1	$+255/256\ V_{in}$
Full scale positive $-$ 1LSB	1	1	1	1	1	1	1	0	$+253/256\ V_{in}$
⋮									
Zero scale (+)	1	0	0	0	0	0	0	0	$+1/256\ V_{in}$
Zero scale ($-$)	0	1	1	1	1	1	1	1	$-1/256\ V_{in}$
⋮									
Full scale negative $+$ 1LSB	0	0	0	0	0	0	0	1	$-253/256\ V_{in}$
Full scale negative	0	0	0	0	0	0	0	0	$-255/256\ V_{in}$

*R_2 matched

Figure 4.8 Two-quadrant multiplication: bipolar digital NO X unipolar analog signal, result in analog form (Courtesy of Precision Monolithics)

necessity of an operational amplifier but the load must provide a path to ground for the common mode output current.

An expression for the differential output current produced by the circuit in Figure 4.9 is readily derived. We have

$$I_1 = I_{o_1} + \overline{I}_{o_2} \quad \text{and} \quad I_2 = \overline{I}_{o_1} + I_{o_2}$$

where $\overline{I}_{o_1} = 225/256$, $I_{ref_1} - I_{o_1}$ and $\overline{I}_{o_2} = 255/256\ I_{ref_2} - I_{o_2}$.

Substitution gives

$$I_1 - I_2 = 2\left(I_{o_1} - I_{o_2}\right) - \frac{255}{256}\left(I_{ref_1} - I_{ref_2}\right)$$

Now

$$I_{o_1} = I_{ref_1}\left(x_1 2^{-1} + x_2 2^{-2} + \ldots + x_8 2^{-8}\right)$$

and

$$I_{o_2} = I_{ref_2}\left(x_1 2^{-1} + x_2 2^{-2} + \ldots + x_8 2^{-8}\right)$$

Thus

$$I_1 - I_2 = \left[2\left(x_1 2^{-1} + x_2 2^{-2} + \ldots + x_8 2^{-8}\right) - \frac{255}{256}\right]$$
$$\left(I_{ref_1} - I_{ref_2}\right) \quad (4.12)$$

But

$$I_{ref_1} = \frac{V_{in}}{R_1} + \frac{V_s}{R_{ref}} \quad \text{and} \quad I_{ref_2} = \frac{V_s}{R_{ref}}$$

Substitution gives

$$I_1 - I_2 = \frac{V_{in}}{R_1}\left[2\left(x_1 2^{-1} + x_2 2^{-2} + \ldots + x_8 2^{-8}\right) - \frac{255}{256}\right] \quad (4.13)$$

Note that the polarity of the differential output current reflects both the polarity of the analog input signal V_{in} and the polarity of symmetrical offset binary bipolar digital code.

DAC–08 s can be used with a variety of different reference amplifier connections (see figure 2.8). The four-quadrant DAC multiplier system of figure 4.9 can be given a high input impedance by using the high impedance bipolar reference configuration. The analog input signal in figure 4.9, instead of being applied through R_1 to the reference amplifier input pin 14, is applied directly to the reference input pin 15 of the upper DAC (DAC 1).

The DAC–08 reference amplifier is frequency compensated by the capacitor C_c which must be connected between pin 16 and the negative supply. In multiplier applications the reference amplifier bandwidth is determined by the value used for C_c – the larger C_c the smaller the bandwidth. The value required for proper frequency compensation depends on the effective source resistance which is connected to pin 14. The minimum recommended values of C_c for adequate phase margin are 15 pF, 37 pF and 75 pF, for source resistance values 1 kΩ, 2 kΩ and 5 kΩ respectively. Larger values of source resistance require proportionately bigger values of C_c.

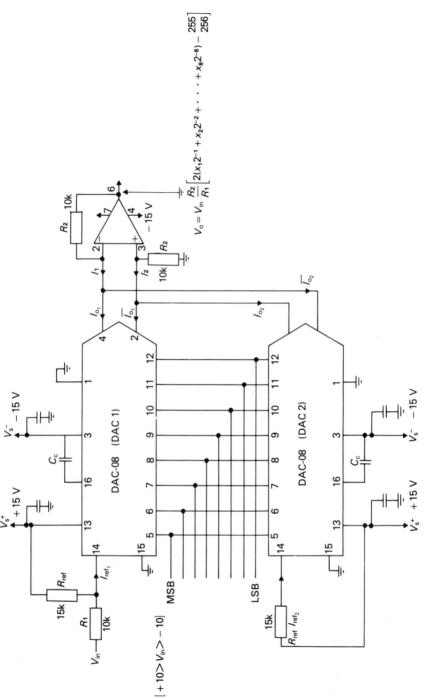

Figure 4.9 Four-quadrant multiplication: bipolar digital NO X bipolar analog signal, result

Digital to Analog Converter Applications

The highest speed operation in multiplier applications of the DAC–08 is obobtained by using a low effective source resistance (200 Ω) at pin 14 and omitting the frequency compensating capacitor entirely. A four-quadrant multiplier DAC system which uses this method of connection for the DAC–08 is shown in figure 4.10. [10] The DAC reference amplifiers maintain a virtual earth at their pin 14 inputs and DAC reference currents are

$$I_{ref_1} = \frac{V_1}{R_{in}} + \frac{V_{ref}}{R_{ref}} \text{ and } I_{ref_2} = \frac{V_2}{R_{in}} + \frac{V_{ref}}{R_{ref}}$$

The differential output current is determined by the same considerations as for the system in figure 4.9. The equation for the differential output current as previously derived in equation 4.12 is

$$I_1 - I_2 = \left[2\left(x_1 2^{-1} + x_2 2^{-2} + \ldots + x_8 2^{-8}\right) - \frac{255}{256} \right] \left(I_{ref_1} - I_{ref_2}\right)$$

Substitution for I_{ref_1} and I_{ref_2} gives

$$I_1 - I_2 = \frac{(V_1 - V_2)}{R_{in}} \left[2\left(x_1 2^{-1} + x_2 2^{-2} + \ldots + x_8 2^{-8}\right) - \frac{255}{256} \right]$$

(4.14)

The system accepts a differential input voltage, $V_1 - V_2$, and produces a differential output current, $I_1 - I_2$, which can be used to drive balanced loads directly. A single ended output voltage can be obtained by adding an operational amplifier configured as a different input current to voltage converter. The output is determined by the product of the differential input and the bipolar, symmetrical offset binary coded digital word which is applied to the DAC's digital inputs. Multiplier operation is in all four quadrants; the output polarity reflects both the analog and digital input polarities. A single ended input signal can be applied to one input with the other input point of the system connected to ground. The input differential and common mode voltage ranges can be extended simply by increasing the value of the resistor R_{in}. Note that the 250 Ω resistors (connected to pin 14) allow the DAC reference amplifier frequency compensating capacitors to be omitted for highest speed operation. If high speed operation is not required the 250 Ω resistors can be removed and frequency compensating capacitors must then be connected between pins 16 and the negative supply.

Bipolar output current multiplying DACs (for example, AD7520 series) are readily adapted to perform a four-quadrant multiplier operation. [14] Suitable circuit connections are given in figure 4.11. The circuit should be compared with the digitally programmed circuit given in figure 4.6. In figure 4.11 a second operational amplifier, A_2, configured as a current inverter has been added. [1] A_2 maintains the I_{o_2} output line of the DAC at ground potential, it

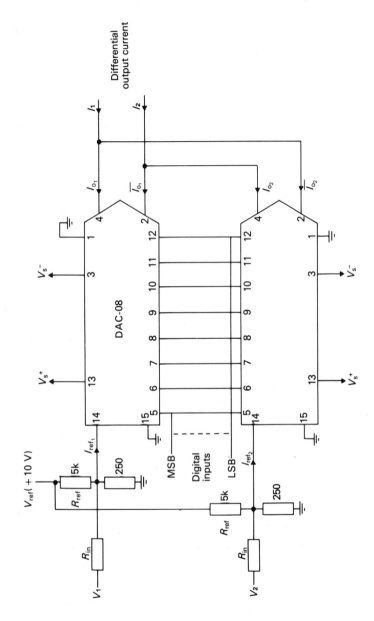

Figure 4.10 High speed four-quadrant multiplier (Courtesy of Precision Monolithics)

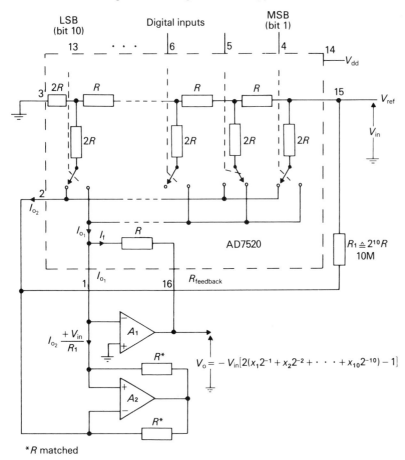

Figure 4.11 Four-quadrant multiplication: bipolar analog X bipolar digitals (Courtesy of Analog Devices)

inverts I_{o_2} and subtracts it from the current flowing through A_1's feedback resistor. An LSB current increment is subtracted from A_1's feedback current via resistor R_1 ($R_1 \triangleq 2^{10} R$). This sets an offset binary code for the DAC's digital inputs (as distinct from the symmetrical offset binary code which would apply if R_1 were omitted).

We derive the ideal performance equation for the circuit as follows. With the I_{o_1} and I_{o_2} output pins maintained at virtual ground the DAC output currents are

$$I_{o_1} = \frac{V_{in}}{R} \left(x_1 2^{-1} + x_2 2^{-2} + \ldots + x_{10} 2^{-10} \right)$$

$$I_{o_2} = \frac{V_{in}}{R} \left(\bar{x}_1 2^{-1} + \bar{x}_2 2^{-2} + \ldots + \bar{x}_{10} 2^{-10} \right)$$

Thus

$$I_{o_2} = (1 - 2^{-10})\frac{V_{in}}{R} - I_{o_1}$$

The current which flows through $\bar{A_1}$'s feedback resistor is

$$I_f = (I_{o_1} - I_{o_2}) - \frac{V_{in}}{2^{10}R}$$

A_1 thus develops an output voltage

$$V_o = -I_f R$$

Substitution for I_f gives

$$V_o = -V_{in}\underbrace{\left[2\left(x_1 2^{-1} + x_2 2^{-2} + \ldots + x_{10} 2^{-10}\right) - 1\right]}_{\text{value of offset binary coded digital input}} \quad (4.15)$$

Operation is in all four-quadrants. The polarity of the analog output is controlled by both the polarity of the analog input and the polarity of the bipolar coded digital input.

4.3.2 Addition and Subtraction of Digital Variables with the Result in Analog Form

The addition and subtraction of digital numbers is traditionally performed using digital ICs. When the results of such operations are to be presented in analog form, savings in the number of IC packages required can sometimes be realised by using analog circuitry to perform the arithmetic operations. If the analog outputs of two DACs are summed the result is an analog signal which represents the sum of the digital numbers which are applied to the DAC's digital input. Subtraction of DAC analog outputs gives a result representing the difference between the digital numbers applied to the DAC's digital input. If current output DACs are used, a single output operational amplifier, suitably configured, can perform the required analog operations.

A basic summing arrangement is illustrated in figure 4.12a. The operational amplifier maintains a virtual ground at its inverting input and thus no demands are placed on the DAC's output voltage compliance limits. The arrangement gives a unipolar positive output signal; the output voltage is related to the digital input words A and B by the equation

$$V_o = I_{ref} R \left[\left(x_{1_A} 2^{-1} + x_{2_A} 2^{-2} + \ldots + x_{n_A} 2^{-n}\right) + \right.$$

Digital to Analog Converter Applications

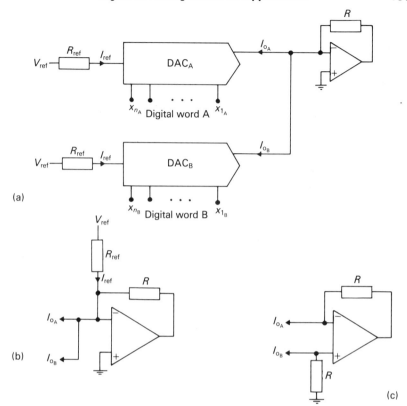

Figure 4.12 Addition and subtraction of digital words with result in analog form: (a) addition of digital words with result in analog form (unipolar); (b) operational amplifier connections for bipolar output; (c) operational amplifier connections for subtraction of digital words

$$\left(x_{1_B} 2^{-1} + x_{2_B} 2^{-2} + \ldots + x_{n_B} 2^{-n} \right) \Big] \tag{4.16}$$

x_{i_A} and x_{i_B} represent the bit values of the natural binary coded digital words A and B respectively.

A bipolar output voltage with an offset binary bipolar code assigned to both words A and B can be obtained by applying an offset current I_{ref} to the summing point of the operational amplifier. The arrangement is illustrated in figure 4.12b, with the offset current applied the analog output voltage being related to the digital words A and B by the equation

$$V_o = I_{ref} R \left[\left(x_{1_A} 2^{-1} + x_{2_A} 2^{-2} + \ldots + x_{n_A} 2^{-n} \right) + \left(x_{1_B} 2^{-1} + x_{2_B} 2^{-2} + \ldots + x_{n_B} 2^{-n} \right) - 1 \right] \tag{4.17}$$

The output represents the algebraic sum of the offset binary coded words A and B in analog form and in all four-quadrants.

Subtraction is accomplished using the arrangement shown in figure 4.12c. The arrangement requires the use of high output voltage compliance DACs (for example DAC–08s). The DAC's output voltage compliance must accommodate their common mode output voltage swing $I_{oB}R$. The output voltage produced by the operational amplifier in figure 4.12c is related to the DAC's digital input words A and B by the equation

$$V_o = I_{ref}R \left[\left(x_{1_A}2^{-1} + x_{2_A}2^{-2} + \ldots + x_{n_A}2^{-n} \right) - \left(x_{1_B}2^{-1} + x_{2_B}2^{-2} + \ldots + x_{n_B}2^{-n} \right) \right] \quad (4.18)$$

A bipolar output signal is produced without the necessity for offsetting the DAC's output. Offsets added to each DAC's output current would be cancelled by the subtraction process. With the offset binary code assigned to words A and B the output gives an analog representation of the algebraic difference between words A and B (A − B) in all four-quadrants.

4.3.3 Digital Multiplication with Result in Analog Form

A multiplying DAC with a variable reference derived from the analog output of a second fixed reference DAC gives an output which is an analog representation of the product of the digital numbers applied to the two DACs. A basic arrangement is illustrated in figure 4.13. It gives a positive unipolar output which represents the product of the unipolar digital numbers A and B expressed in analog form.

An expression for the output voltage produced by the circuit in figure 4.13 is readily derived. The DAC labelled B has a variable reference controlled by the output current of the DAC labelled A.

$$I_{ref_B} = \frac{I_{o_A} R_2}{R_3}$$

where

$$I_{o_A} = I_{ref_A} \left(x_{1_A}2^{-1} + x_{2_A}2^{-2} + \ldots + x_{8_A}2^{-8} \right)$$

The output voltage is

$$V_o = I_{o_B} R$$

where

$$I_{o_B} = I_{ref_B} \left(x_{1_B}2^{-1} + x_{2_B}2^{-2} + \ldots + x_{8_B}2^{-8} \right)$$

Algebraic manipulation yields the final performance equation as

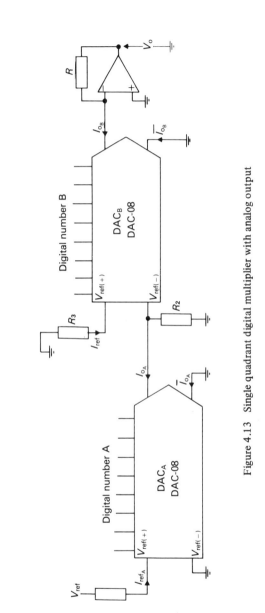

Figure 4.13 Single quadrant digital multiplier with analog output

$$V_o = \frac{R_2 R}{R_3} I_{\text{ref A}} \left[\left(x_{1_A} 2^{-1} + x_{2_A} 2^{-2} + \ldots + x_{8_A} 2^{-8} \right) \right.$$
$$\left. \times \left(x_{1_B} 2^{-1} + x_{2_B} 2^{-2} + \ldots + x_{8_B} 2^{-8} \right) \right] \quad (4.19)$$

a single quadrant digital multiplication with analog output. x_{i_A} and x_{i_B} represent the bit values of the natural binary digital numbers A and B respectively.

Two-quadrant multiplication in which a bipolar, symmetrical offset binary code is assigned to variable B can be easily arranged. The output operational amplifier is simply configured as a differential input current to voltage converter to produce an output voltage determined by the difference between the normal and complementary current outputs of DAC$_B$.

A four-quadrant digital multiplier which uses three DACs is shown in figure 4.14. [10] The arrangement consists essentially of a third DAC added to the multiplying DAC system given in figure 4.10. DAC 3 supplies a variable analog reference to DACs 1 and 2.

The equation for the differential output current produced by the DAC 1 and 2 combination is as derived previously (see equation 4.12).

$$I_1 - I_2 = \left[2 \left(x_{1_B} 2^{-1} + x_{2_B} 2^{-2} + \ldots + x_{8_B} 2^{-8} \right) - \frac{255}{256} \right]$$
$$\left(I_{\text{ref}_1} - I_{\text{ref}_2} \right)$$

I_{ref_1} and I_{ref_2}, controlled by DAC 3's output current are

$$I_{\text{ref}_1} = \frac{V_{\text{ref}}}{R_{\text{ref}}} - \bar{I}_{o3} \text{ and } I_{\text{ref}_2} = \frac{V_{\text{ref}}}{R_{\text{ref}}} - I_{o3}$$

where

$$I_{o3} = \frac{V_{\text{ref}}}{R_{\text{ref}_3}} \left(x_{1_A} 2^{-1} + x_{2_A} 2^{-2} + \ldots + x_{8_A} 2^{-8} \right)$$

and $\bar{I}_{o3} = \frac{255}{256} \frac{V_{\text{ref}}}{R_{\text{ref}_3}} - I_{o3}$

Algebraic manipulation gives the circuit performance equation as

symmetrical offset
binary number A

$$I_1 - I_2 = \frac{V_{\text{ref}}}{R_{\text{ref}_3}} \left[2 \left(x_{1_A} 2^{-1} + x_{2_A} 2^{-2} + \ldots + x_{8_A} 2^{-8} \right) \right] - \frac{255}{256}$$

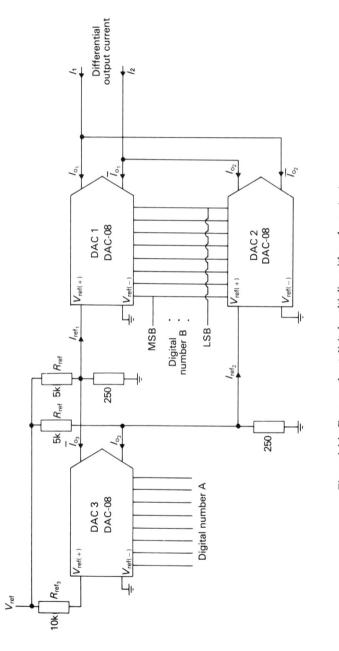

Figure 4.14 Four-quadrant digital multiplier with analog output

symmetrical offset binary number B

$$\times \left[2 \left(x_{1_B} 2^{-1} + x_{2_B} 2^{-2} + \ldots + x_{8_B} 2^{-8} \right) - \frac{255}{256} \right] \quad (4.20)$$

The differential output current can be used to drive a balanced load directly or it can be converted to a single ended output voltage as described previously.

4.4 DIGITAL SCALE SETTING (OFFSET ZEROING)

Many analog systems require a means of output voltage adjustment. For example, it may be necessary to set the output to some calibrated value prior to a test procedure or it may be necessary to null the input offset voltage of an operational amplifier. Potentiometers are the circuit elements normally used in manual adjustment procedures. Multiplying DACs can in a sense be regarded as digitally controlled potentiometers (figure 4.15); as such, they can sometimes replace

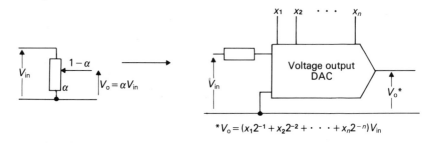

Figure 4.15 Multiplying DAC acts as digitally controlled potentiometer

potentiometers so as to provide an automatic digital scale setting facility. The principles involved in this type of application are illustrated in figure 4.16. [17]

The output voltage produced by the lower DAC in figure 4.16 serves as the adjustable input control signal (the input offset) to the controlled analog system. The DAC's output voltage is incremented one LSB at a time, either up or down, depending on the count mode of the binary counter which is used to supply its digital input signal. A comparator senses whether the output of the controlled system is greater or less than its required value and adjusts the count mode accordingly. The system acts as a tracking converter with the controlled analog system forming part of the tracking converter's feedback loop. When the controlled system's desired output is reached, the digital input to the control DAC 'hunts' between the two values which straddle the required setting. The input control signal (the input offset) is thus set to an accuracy ± 1 LSB. The signal is 'held' when a disable command is applied to the counter. DAC 1 works automatically to set the output of the analog system to the comparator reference voltage level. This voltage may be zero or a calibrating voltage which can be established automatically by DAC 2.

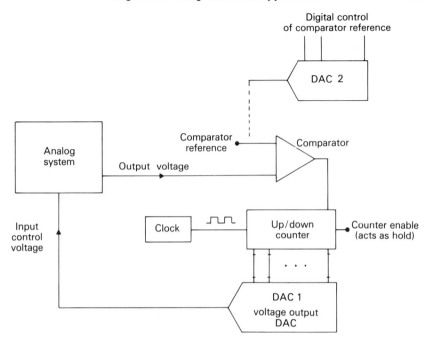

Figure 4.16 Digital output voltage adjustment for analog system

Systems similar to the above can be used for the automatic offset nulling of operational amplifiers. Some operational amplifier types can have their normal offset nulling potentiometer directly substituted by a current output DAC as shown in figure 4.17. [18] In this circuit the complementary current outputs of the DAC–08 unbalance the operational amplifier differential input stage collector currents allowing the amplifier input offset voltage to be nulled.

4.5 DIGITAL CONTROL OF FREQUENCY

There are a number of monolithic integrated circuit devices available for the generation of analog waveforms; some of these readily permit a voltage control of frequency. [9] There are also voltage to frequency converters available in both monolithic and modular form; the specific function of these components is to produce a pulse train with frequency proportional to an analog control voltage. Any of these systems which are capable of producing a waveform with frequency controlled by an analog voltage, can be readily adapted to a digital control of frequency. It is simply a matter of using a DAC to generate the analog control voltage. The control voltage produced by the DAC and hence the frequency of oscillations of the waveform generator are then directly controlled by the digital input to the DAC. Other circuit configurations for digital control of frequency can be devised in which the DAC forms a more integral part of the waveforming system.

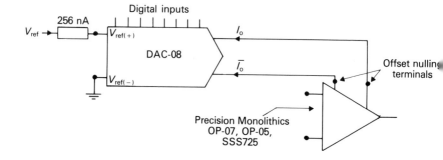

Figure 4.17 Digitally controlled offset nulling of operational amplifier

In the system shown in figure 4.18 a current output DAC gives a digital control of frequency capability to the well-known operational amplifier integrator-regenerative comparator waveform generation. [1] Transistors Q_1, Q_2 form a current mirror; any collector current established in Q_1 by the DAC's output current flows as collector current in Q_2 towards the integrator summing point. When the output of the comparator A_2 is at its negative saturation limit the current mirror is disabled by diode D_2 and the DAC output current flows away from the integrator summing point through diode D_1. When the comparator output is at its positive saturation limit, D_2 and the current mirror conduct. D_1 is reverse biased and the collector current of Q_2, which is equal to the DAC output current, flows towards the integrator summing point.

The output of the integrator thus ramps up when the comparator output is negative and ramps down when the comparator output is positive. A triangular waveform appears at the output of the integrator, its positive and negative peak values going between the input threshold values of the regenerative comparator.

The peak to peak size of the triangular waveform is

$$\left(\left| V_{o\,\text{sat}}^- \right| + V_{o\,\text{sat}}^+ \right) \frac{R_1}{R_2}$$

If the comparator positive and negative saturation input limits are equal in magnitude the triangular wave has a peak-to-peak value

$$2 V_{o\,\text{sat}} \frac{R_1}{R_2}$$

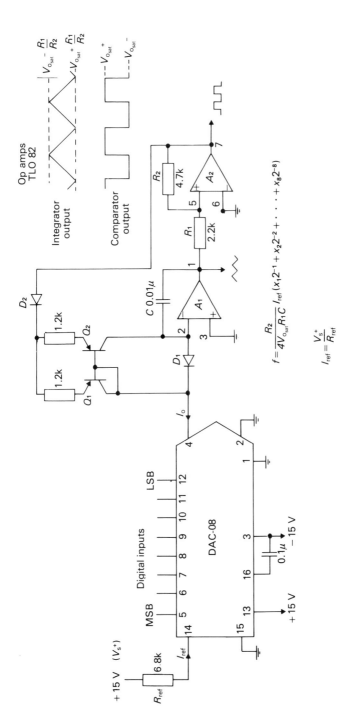

Figure 4.18 Digital control of triangular-square wave generator

The time taken for the integrator output to ramp between its limit values is

$$T_1 = T_2 = \frac{2V_{o\,sat}\, R_1/R_2}{I_o/C} \quad \text{s}$$

And the frequency of oscillations is thus

$$f = \frac{1}{T_1 + T_2} = \frac{R_2}{4V_{o\,sat}\, R_1 C}\, I_o$$

But the DAC output current I_o is determined by the digital inputs applied to the DAC. Thus

$$f = \frac{R_2}{4V_{o\,sat}\, R_1 C}\, I_{ref}\left(x_1 2^{-1} + x_2 2^{-2} + \ldots + x_n 2^{-n}\right) \qquad (4.21)$$

Frequency stability depends on the stability of the DAC reference current I_{ref} and the operational amplifier's output saturation limit $V_{o\,sat}$. Both I_{ref} and $V_{o\,sat}$ depend on power supply voltage and the the influence of the power supply voltage on frequency is therefore partially cancelled out.

Our analysis of the circuit has been based on the assumption of ideal operational amplifier characteristics. In the practical circuit the slew rate of the operational amplifier limits the rise and fall times of the comparator output square wave. This effect limits the upper operating frequency of the circuit. Integrator drift sets the lower operating frequency limit.

Any mismatch in transistors Q_1 and Q_2 makes the capacitor charge and discharge currents unequal and produces an asymmetrical waveform. Emitter resistors can be trimmed for waveform symmetry. Alternatively a variable emitter resistor can be used to act as a waveform symmetry control.

The wide output voltage compliance of current output DACs like the DAC–08 makes it possible directly to charge a timing capacitor with the DAC output current. A circuit which uses a DAC in this way (Precision Monolithics Companding DAC–76) to give an exponential digital control of frequency is described in reference 19. In the system described, comparators, referenced by a resistive divider connected across the supply rails, are used to fix the upper and lower capacitor charging limits.

The system shown in figure 4.19, consisting of a DAC–08, a current mirror and a 555 timer, is another digitally controlled oscillator. The 555 timer contains two comparators, the so-called threshold and trigger comparators. Comparator reference levels are derived from a resistive divider across the 555 supply rails. The upper threshold level is set at $\frac{2}{3}V_s$ and the lower trigger level is set at $\frac{1}{3}V_s$. In figure 4.19 the timing capacitor C is charged linearly up and down between these limits. The current mirror formed by transistors Q_1 and Q_2 functions in a similar way to that shown in the circuit of figure 4.18.

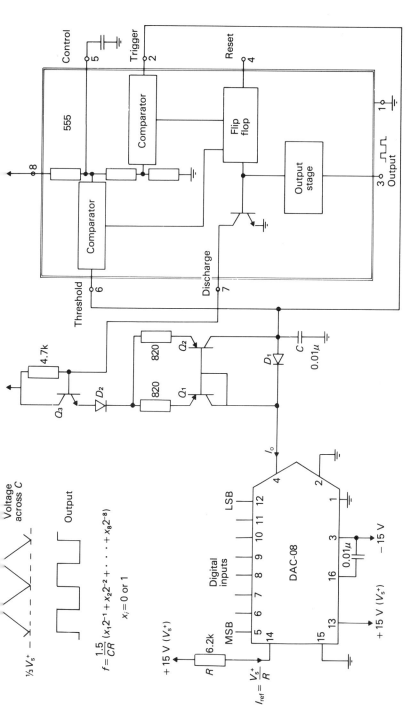

Figure 4.19 Digitally controlled oscillator

The DAC's output current, inverted by the current mirror, causes a linear charging of the timing capacitor C; diode D_1 is reverse biassed during this part of the cycle. When the voltage across the capacitor reaches the threshold level $\frac{2}{3}V_s^+$ the discharge transistor in the 555 timer is switched on. This reverse biases diode D_2 and prevents conduction through the current mirror transistors Q_1 and Q_2. The diode D_1 now conducts, the DAC's output current flows through D_1 and C charges down linearly. When the voltage across C reaches the trigger level $\frac{1}{3}V_s^+$, the 555 output goes high, the discharge transistor in the 555 is switched off and the capacitor is again charged by the current mirror.

It is essential to buffer the capacitor voltage if the triangular wave is to be applied to an external load. If this is not done load current will subtract from capacitor charging current and influence timing periods. A follower connected FET input operational amplifier (not shown in figure 4.19) can be used to buffer the capacitor voltage. The buffer should be inserted between the capacitor and the timer where it will overcome errors due to the 555 comparator input currents.

Assuming that capacitor charge and discharge times are equal the period of the oscillations produced by the circuit is

$$T = 2 \frac{\frac{1}{3}V_s^+}{I_o/C} \text{ s}$$

where $I_o = I_{ref}(x_1 2^{-1} + x_2 2^{-2} + \ldots + x_8 2^{-8})$ and $I_{ref} = V_s^+/R$. Substitution gives the expression for the frequency of the oscillations as

$$f = \frac{1}{T} = \frac{1.5}{CR}\left(x_1 2^{-1} + x_2 2^{-2} + \ldots + x_8 2^{-8}\right) \tag{4.22}$$

4.6 DIGITAL CONTROL OF A LINEAR RAMP

A 555 timer used in the monostable mode produces a linear ramp when the timing capacitor is charged by a constant current. [9] In figure 4.20 a current mirror driven by a current output DAC is used to provide linear capacitor charging. It gives a digital control of both ramp rate and monostable timing period to the 555 circuit. The ramp rate is

$$\frac{dV_c}{dt} = \frac{I_o}{C} \text{ V/s}$$

where

$$I_o = \frac{V_s^+}{R}\left(x_1 2^{-1} + x_2 2^{-2} + \ldots + x_8 2^{-8}\right)$$

The monostable timing period is

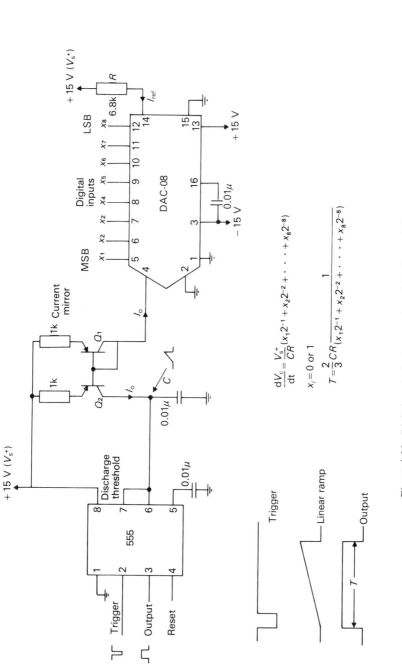

Figure 4.20 Digital control of ramp rate and timing period

$$T = \frac{\frac{2}{3} V_s^+}{dV_c/dt}$$

$$T = \frac{2}{3} CR \left(\frac{1}{x_1 2^{-1} + x_2 2^{-2} + \ldots + x_8 2^{-8}} \right) \qquad (4.23)$$

The ramp slope is directly proportional to the digital input number applied to the DAC. The timing period is inversely proportional to that number. Note that the timing period is independent of the supply voltage value.

A follower connected FET input operational amplifier should be used to buffer the capacitor voltage if the ramp is used to drive an external load.

4.7 GENERATION OF FUNCTIONAL RELATIONSHIPS

In analog signal processing and conditioning systems signals are sometimes subjected to non-linear processing operations. A circuit used to perform non-linear processing is often referred to as a function circuit. Function circuits are used to give defined mathematical functional relationships between input and output signals such as, $V_o = V_{in}^2$, $V_o = \sin V_{in}$, $V_o = \log V_{in}$, etc. Sometimes purely arbitarary functional relationships are required, for example, in linearising the output of a non-linear transducer.

Analog function circuits can be made using operational amplifier based circuitry [1, 23, 24] ; such circuits can be bought as ready built function modules (see Analog Devices or Burr Brown product guides). However, with the decreasing cost of monolithic data converter and semiconductor memories a digital approach to the generation of analog functional relationships is finding increasing acceptance.

Semiconductor read only memories, ROMs, can permanently store large amounts of digitally coded data, their storage capacity being designated in terms of number of words and number of bits in each word. A ROM designated as a 512 X 8 device has a storage capacity for 512 8-bit words. When a ROM is used in generation of analog functional relationships, values of the function are not really 'generated' in the sense of the continuous generation performed by analog circuitry. Discrete values of the function are permanently stored in digital form in memory. The values are routed, one at a time, to the memory output in response to a coded digital word applied to the memory input (the memory address code).

A system is outlined in figure 4.21. Analog input signals are converted into digital form by an A/D converter. The digital output of the converter acts as the memory address word causing the digital value of the required function to appear at the memory output. An output DAC converts the memory output into analog form. In a sense the system performs a table look-up operation, the ROM behaving as a table of numerical values. For example, in the case of a ROM designed to store the trigonometrical sine function its input address codes are

Digital to Analog Converter Applications

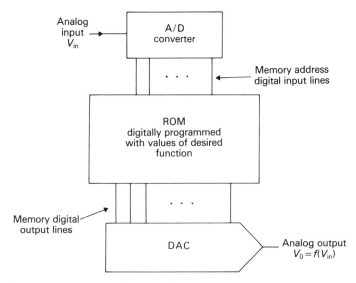

Figure 4.21 Data converter and ROM used in 'generation' of analog function

assigned to represent angles between 0° and 90° and stored at the memory locations corresponding to the addresses are the digitally encoded values of the sines of the angles (see exercise 8).

The functional relationship that a ROM is required to simulate must be specified at the time it is purchased — it cannot be changed thereafter. An alternative is to buy a programmable read only memory, a PROM; such a device can be programmed by the user to simulate any functional relationship he requires. It should be remembered that in all digital representations of analog quantities there is an inherent quantisation uncertainty limiting resolution. Clearly ROM storage capacity sets a limit to the resolution and accuracy obtainable with the digital techniques of function generation.

4.8 DIGITAL GENERATION OF ANALOG WAVEFORMS

A DAC's ability to change digital data into analog form can be used as a basis for generating time varying analog waveforms from time varying digital data. The principle underlying a digital generation of analog waveforms is illustrated by the arrangement shown in figure 4.22. The system consists of a digital clock, a digital counter and a DAC.

Clock pulses increment the logic states of the counter output lines through a binary counting sequence. The counter outputs provide the DAC's digital inputs and the sequential counter steps cause a sequential incrementing of the DAC's analog output.

If the counter is in the count up mode its digital outputs sequence from all '0's up to '1's. When the counter is full (all '1's) it returns to its empty state (all

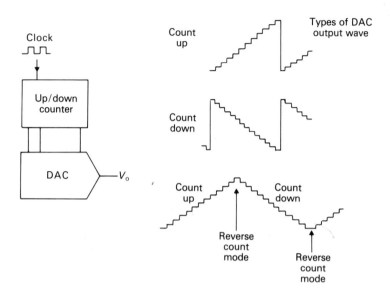

Figure 4.22 Digital waveform generation

'0's) and starts counting up again. The analog output produced by the DAC is a positive going sawtooth. Setting the counter in the count down mode causes it to sequence down form all '1's to all '0's and the DAC produces a negative going sawtooth. A triangular wave is produced if it is arranged that the counter changes counting directions when it is full (all '1's) and again when it is empty (all '0's).

A practical waveform generating circuit using the DAC-08 is shown in figure 4.23. The system includes logic circuitry to control counting mode and allow a selection between positive going sawtooth, negative going sawtooth and triangular waves. Appropriate switching of the DAC–08's I_o and $\overline{I}_o$ outputs to the operational amplifier current to voltage converter gives unipolar positive, unipolar negative or bipolar output waveforms. The system can be modified to produce trapezoidal waveforms by additional logic circuitry used to disable the counter for a specific number of clock pulses before count reversal.

Extending a DAC digital counter waveform generator system to include a ROM allows other repetitive waveforms to be generated. A ROM which is digitally programmed with the desired function is used. The counter output acts as the ROM memory address causing the digital words stored at the separate memory locations to appear in time sequence at the memory output. An output DAC converts the digital data into analog form.

Digital to Analog Converter Applications

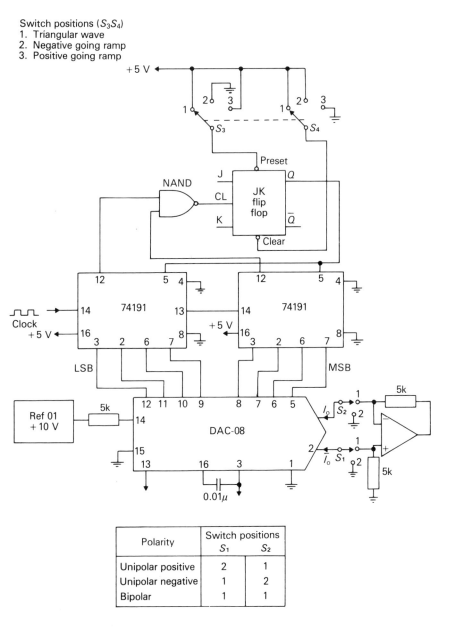

Figure 4.23 Practical waveform generator circuit

4.9 SELF-ASSESSMENT EXERCISES

4.9 EXERCISES

1. An outline diagram for a manually controlled digitally programmed voltage source is given in figure 4.24. The arrangement is intended to allow voltages $0.0, 0.1, 0.2, \ldots, 9.8, 9.9$ to be set by the adjustment of two thumbwheel switches. The switches are each four-pole, ten-way devices. Find the value required for R_1 and indicate the additional wiring which is required to obtain the correct BCD coding. A switch point connected to ground applies logical '0', a switch point connected to +5 V applies logical '1'.

2. Indicate the modifications which must be made to the circuit in Figure 4.24 in order to change the system into a digitally programmed current source (see figure 4.3). The system is to be used to switch currents $0.0, 0.1, 0.2, \ldots, 9.8, 9.9$ mA into a grounded resistive load. Estimate the maximum value of this resistive load for which the system's output voltage compliance limit is not exceeded.

3. The following values are used in the circuit of figure 4.5: $V_s = +12$ V, $R_{ref} = 6$ kΩ, $R_1 = 5$ kΩ, $R_2 = 10$ kΩ. If two 6-bit DACs are used what is the gain of the circuit when the digital inputs: (a) 100100, (b) 011011, are applied? What restriction must be placed upon the range of V_{in}? Explain the reasons for this restriction.

4. What digital bit values must be applied to the DAC shown in figure 4.7 in order to set the gain of the circuit at: (a) -4, (b) -32, (c) -256?

5. In figure 4.18 indicate the path of the DAC's output current and the direction of the capacitor charging current when the comparator output is (a) high, (b) low.

6. With the component values as shown in figure 4.18 find
    (a) the value of the integrating capacitor charge and discharge currents,
    (b) the slope of the triangular wave in V/s,
    (c) the amplitude of the triangular wave,
    (d) the frequency of oscillations.
The digital input 01000000 is applied to the DAC. Assume that amplifier A_2 has output saturation limits ± 10 V, and that circuit components behave ideally.

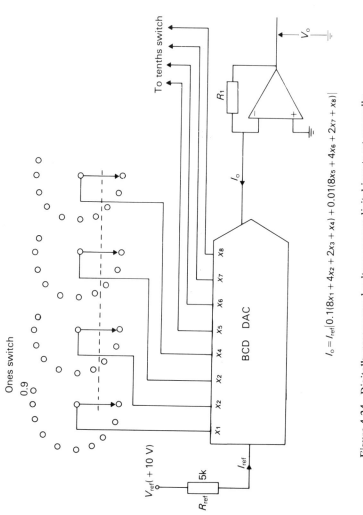

Figure 4.24 Digitally programmed voltage source; digital inputs set manually

Data Converters

7. Repeat question 6 for a circuit in which the emitter resistors of Q_1 and Q_2 are 1 kΩ and 2 kΩ respectively.

8. A system of the type shown in figure 4.21 is to be used as a means of generating the functional relationship $V_o = \sin V_{in}$. Values of the output voltage in the range 0 to 10 V are to represent angles in the range 0° to 90°. The system is to use a 32 × 8-bit ROM with the ROM's 4-bit memory address supplied by a 4-bit A/D converter and an 8-bit DAC used to produce the analog output voltage. Find the various scaling factors which must be used in the system in order that a 10 V maximum input signal should produce a 10 V maximum output. Find the bit values which must be programmed into the ROM. Give a graph showing the form of the output/input relationship which you expect the system to generate.

9. Roughly sketch the time variations of the output voltage which you expect to obtain in the system of question 8. If the ROM's memory address is incremented by:
    (a) a continuously clocked 4-bit binary up counter,
    (b) an up/down counter whose count mode 'changes when it is full (all '1's) and when it is empty (all '0's)'.
Suggest possible modifications to the system which would cause it to produce a sinusoidal output wave. Say how you expect the frequency of this sine wave to relate to the system's clock frequency.

10. A 1 mHz clock signal is applied to the waveform generator system shown in Figure 4.23. Sketch the waveforms which you expect the system to generate for the following switch settings. Indicate wave amplitude, polarity and frequency.
    (a) S_3, S_4 position 1; $S_1 S_2$ position 1.
    (b) S_3, S_4 position 1; S_1 position 1, S_2 position 2.
    (c) S_3, S_4 position 3; S_1 position 2, S_2 position 1.

5 Digital Processing of Analog Signals

Analog signals invariably suffer some degradation when they are transmitted or processed using analog circuit techniques. Analog circuits pick up noise and are subject to errors. Digital circuit techniques are not so error prone—they allow the error free transmission and processing of information contained in a digital signal. Data converters make it possible to gain the advantages of digital processing techniques in performing analog functional operations. An A/D converter can convert an analog signal into digital form; digital ICs can then be used to perform processing operations and a DAC used to turn the result back into analog form.

Analog signals have a continuous range of values open to them and they vary continuously with time. In contrast, a digital variable can only take on a series of discrete values. A specific digital number can represent only an instantaneous value of an analog variable. A digital representation of the time variations of an analog signal is in the form of a set of digital numbers representing instantaneous values of the analog signal at equally spaced times. Conversion of the time variations of an analog signal into digital form involves taking a series of instantaneous samples of the analog signal. Provided that the samples are taken frequently enough a theorem called the *sampling theorem* tells us that all the information contained in the analog signal is also contained in its samples.

5.1 THE SAMPLING THEOREM

The sampling theorem states

> A continuous signal can be represented by, and reconstructed from, a set of instantaneous measurements or samples of the signal which are made at equally spaced times. The signal must be sampled at a frequency which is greater than twice the highest frequency component present in the signal.

A mathematical proof of the sampling theorem will not be attempted. What the theorem means in relationship to sine waves can be understood from figure 5.1. Voltage sine waves of different frequencies are shown. A series of samples of the waves, taken at equally spaced times, are represented by the vertical lines shown superimposed on the sine waves. The time interval between samples is T_s, the sampling frequency $f_s = 1/T_s$ and f is the frequency of the sine wave.

In figures 5.1a and b the sampling frequency is greater than twice the frequency of the sine wave. In each case there is only one sinusoid that can be drawn through the sample values—the one shown in the diagram. The samples shown in figures 5.1a and b fully quantify the sine waves.

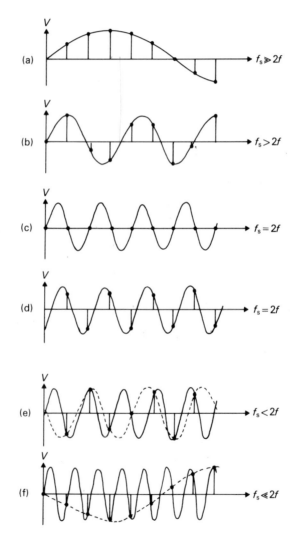

Figure 5.1 Sampling a sine wave. Sampling frequency f_s, sine wave frequency f

Digital Processing of Analog Signals

In figures 5.1c and d the sampling frequency is exactly twice the sine wave frequency. The all zero samples of figure 5.1c clearly provide no information about the sine wave. Sampling instants have been shifted in figure 5.1d. The samples now describe the frequency of the sine wave but do not uniquely quantify its amplitude. A whole series of sine waves of different amplitude but the same frequency could be drawn through the sample values in figure 5.1d. Thus, if the sampling frequency is twice the frequency of a sinusoid the samples do not fully quantify the sine wave.

The sampling frequency in figure 5.1e and f is less than twice the sine wave frequency. In both cases a second sine wave is shown by a broken curve which passes throught the same sample points as those obtained from the original sine wave. Given only the sample values it would not be possible to deduce which of the two sine waves shown has been sampled. The construction of two different sinusoidal signals from a given set of sample values is called *aliasing*. Aliasing leads to ambiguity: *a sine wave must be sampled at a frequency more than twice the frequency of the sine wave in order to avoid ambiguity*.

The amount of information carried by a continuous sine wave is limited by the number of quantities required to specify it, namely its amplitude, frequency and phase. The analog signals encountered in electronic systems generally have a non-sinusodial waveform. However, any signal waveform can be regraded as being made up from its Fourier components, that is, it can be thought of as consisting of a whole series of sine waves of different frequencies. Thus, if a non-sinusoidal analog signal is to be correctly sampled a sampling frequency must be chosen which is more than double the frequency of the highest frequency component of the analog signal. The sampling theorem then tells us that all the information contained in the analog signal is contained in its samples. The analog signal can be transmitted from one place to another in the form of its samples. It can then be reconstructed from its samples.

The reconstruction of an analog signal waveform from its repetitive sample values is accomplished by passing the sequence of samples through a low pass filter. The process is illustrated in figure 5.2. The cut-off frequency of the low pass filter is made equal to half the sampling frequency f_s. The low pass filter then strongly attenuates the signal component at the sampling frequency which is present in the sequence of samples. All the frequency components present in the original signal, which are contained in the sequence of samples, are passed by the filter without significant attenuation. They appear at the filter output as a reconstructed version of the original signal. The reconstructed version of the signal is reduced in amplitude as compared with the amplitude of the samples.

Attenuation and/or phase shift of the frequency components of the original signal with frequency close to the cut-off frequency of the low pass filter can lead to distortion of the reconstructed output wave. In order to minimise such distortion the sampling frequency f_s should be made appreciably bigger than twice the highest frequency component of the original signal. Since the cut-off frequency of the low pass filter is half the sampling frequency the

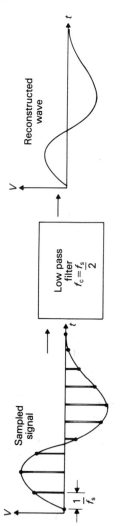

Figure 5.2 Reconstruction of analog signal from its samples with low pass filter

Digital Processing of Analog Signals

highest frequency components of the original signal will then lie well within the bandwidth limit of the low pass filter. The filter will cause negligible attenuation and phase shift of the high frequency components of the signal.

5.2 SAMPLING AND DIGITISATION; QUANTISATION NOISE

The sampling theorem is concerned with the representation of a time varying analog signal by a set of instantaneous analog values or samples of the signal. If digital signal processing techniques are to be used the analog sample values must be converted into digitally encoded sample values. The digitisation process consists of dividing the continuous amplitude range of the sampled analog signal into a fintie number of discrete levels (called *quantisation*). The amplitude of each sample value is then referred to that of the nearest discrete level. The operation is performed by an A/D converter.

Information contained in digitally encoded sample values can be transmitted and processed without the degradations and errors associated with analog signal processing. However, even in ideal sampling and digitisation systems the information contained in the digital sample values cannot faithfully represent the information contained in the analog signal from which the digital samples were derived. The digital samples represent a distorted version of the original analog signal. The distortion is an irremoveable error introduced not by the signal sampling, but by the necessity for quantisation—it is called *quantisation distortion*.

Quantisation distortion is caused by the errors inherent in digital sample values. The error in any particular digital sample (the quantisation error) depends on how close the analog signal amplitude was to a quantised level at the instant that the sample was taken. In the continuous sampling and digitisation of a complex analog signal with amplitude spanning many quantised levels, the sampling instants are in no way synchronised to the frequency components of the analog signal. There is no predictive relationship between the error of one digital sample and that of the next. The quantisation error fluctuates randomly from sample to sample. The maximum fluctuation possible in the quantising error is determined by the separation of the quantising levels. Assuming equally spaced levels (linear A/D conversion) the maximum error fluctuation is equal to the analog value allocated to the LSB in the A/D conversion relationship.

Quantisation distortion is often called *quantisation noise* because of the random nature of the quantisation errors which cause it. When digital sample values are converted back into analog form the reconstructed analog signal has quantisation noise added to it as an unwanted noise component. If the original analog signal is complex and its amplitude spans many quantised levels the quantisation noise power is distributed uniformly throughout the range of frequencies spanned by the analog signal. It cannot therefore be removed by filtering the reconstructed analog signal. It has the properties of an added white noise component.

126 Data Converters

In figure 5.3 we illustrate the sampling and digitisation of an analog signal voltage with a 4-bit binary code used to represent digitial values. The analog range is divided into 15 discrete levels, 16 if we count the zero level. ($16 = 2^n$ where $n = 4$, the number of bits in a binary code). If the analog signal has a

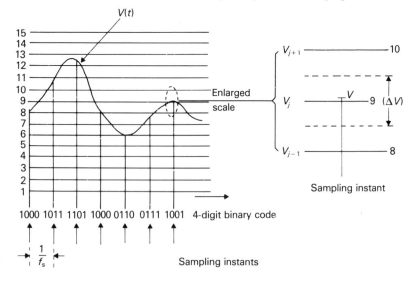

Figure 5.3 Digitisation of analog signal with 4-bit binary code

value v at an instant of sampling and the nearest quantised level is v_j, the quantising error is due to $v - v_j$. A sample taken close to the 9th quantised level is shown in enlarged scale.

Figure 5.3 shows just a small 'time window' of a continuous sampling and digitisation process. In this process every time a sample value occurs within the range $v_j - (\Delta v)_j/2$ to $v_j + (\Delta v)_j/2$ the digital code word corresponding to the jth level is assigned to it. Associating a mean square error voltage $\overline{(v - v_j)^2}$ with the jth level, the mean square value of the quantisation noise associated with the continuous process may be computed as the sum of the mean square error voltages associated with each level. If the level separations are small compared with the total signal amplitude range and if the levels are equally spaced, (linear quantisation $\Delta v_j = \Delta v$ a constant) the computation gives [20]

$$N_Q{}^2 = \frac{(\Delta v)^2}{12} \tag{5.1}$$

where $N_Q = \Delta v/\sqrt{12}$ is the RMS value of the quantisation noise and Δv is the separation of the quantised levels.

$$\Delta v = \frac{\text{full scale analog range}}{2^n}$$

Digital Processing of Analog Signals

where n is the number of bits in the A/D conversion code.

The expression for N_Q may be used to indicate the signal-to-noise ratio, SNR, to be expected when a sinusoidal analog signal is sampled and digitised. We consider a full scale sine wave, that is, one whose peak-to-peak range spans the whole of the analog range of the A/D converter used in the system. The peak value of such a full scale sine wave is $(2^n \Delta v)/2$. Half of the 2^n quantised levels are used for each polarity of the sine wave. The RMS value of a full scale sine wave is

$$V_{RMS\;fullscale} = \frac{2^{n-1} \Delta v}{\sqrt{2}}$$

The ratio of signal power to quantisation noise power is

$$SNR = \frac{V_{RMS\;fullscale}}{N_Q} = \frac{2^{n-1} \Delta v}{\sqrt{2}} \bigg/ \frac{\Delta v}{\sqrt{12}}$$

$$SNR = \sqrt{(1.5)} 2^n \tag{5.2}$$

Expressing the SNR in dBs

$$SNR_{dB} = 20 \log_{10} \sqrt{(1.5)} 2^n = 6.02n + 1.76 \text{ dB} \tag{5.3}$$
(fullscale sinewave)

As is to be expected, the SNR is dependent on the number of quantisation levels used in the digitisation process. The more levels, the smaller the separation of the levels and the smaller the quantisation noise. The number of quantisation levels is 2^n where n is the number of binary bits used in the A/D conversion. In a linear conversion each bit contributes 6 dB to the system performance ($20 \log_{10} 2 \triangleq 6$).

The expression given above for SNR is applicable to a full scale sine wave signal. Sine wave signals with amplitude less than full scale will give a smaller SNR, (reduction in SNR being in proportion to the reduction in sine wave amplitude below full scale). The maintenance of a high SNR for conversion with as many bits as possible. This of course assumes that other performance errors of the practical converter do not exceed its inherent quantisation error.

A 16-bit linear conversion is recommended as the ideal for the digitisation of very high fidelity audio signals. [23] The cost of a 16-bit A/D is many times greater than that of, say, an 8-bit converter. A reduction in audio quality requirements, for example, such as would be tolerable in the music distribution systems used in commercial aircraft (see later) allows the use of converters with a smaller number of bits and keeps costs down. There are specialised conversion techniques which have been devised for use in digital audio systems which attempt to maintain quality with a reduced number of bits; [23] in the digitisation of speech signals a non-linear quantisation is sometimes used. The non-linearity is designed to suit the amplitude distribution of the speech signal. The discrete quantised levels into which the analog range is divided are more closely spaced at the bottom of the range than at the top. The non-linear systems do not suffer the same reduction in SNR with decrease in analog signal amplitude as do linear conversion systems [20, 21]

The principles underlying the sampling and digitisation of analog signals which have been outlined in this section are by no means new. Sampling and digitisation has been in use by the communications industry for many years under the designation *pulse-code modulation*, PCM. An understanding of the principles is necessary if the reader is to appreciate the recent advances in the field. Most recent advances are not in the development of new techniques, but in the application of new low cost IC devices to the implementation of previously established techniques.

5.3 THE PRACTICAL IMPLEMENTATION OF SAMPLING AND DIGITISATION

The main functional elements of a sampling and digitisation system are outlined in figure 5.4. They consist of a low pass filter, a sample/hold module, an A/D converter and digital control circuitry used to direct the system operation. The low pass filter is used to set a limit to the bandwidth of the analog signal. No low pass filter is required if the analog signal is already of defined bandwidth; for example, a bandwidth limit could be imposed by previous conditioning of the analog signal.

The operation of the sample/hold may be modelled in terms of a switch connected to a capacitor. In the sample mode the switch is closed and the voltage across the capacitor follows or tracks the input signal. The voltage voltage across the capacitor constitutes the sample/hold output. When the switch is opened (hold mode), the voltage across the capacitor ideally holds constant at the value it had at the instant the switch opened. Practical sample/holds are subject to a variety of performance limitations because of finite capacitor charging time, capacitor leakage and switch deficiencies. The reader is referred elsewhere for a discussion of practical sample/hold circuits. [1, 3, 4, 5] For our current purposes we shall assume an ideal sample/hold behaviour.

The sample/hold presents a series of instantaneous samples of the analog signal as an input signal to the A/D converter. Sample values are held constant while the converter digitises them. The converter produces a digital word corresponding to each analog sample; the digital data may be required in serial or parallel form. The process is governed by a series of digital control signals in the following sequence.

(1) Command sample/hold to hold.
(2) Instruct A/D converter to start conversion, SC.

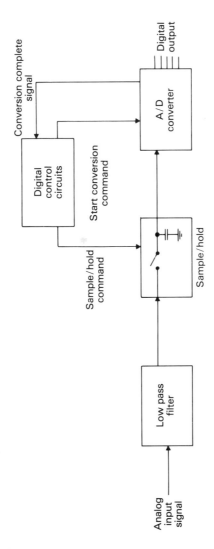

Figure 5.4 Functional elements of sampling and digitising system

(3) A/D converter signals conversion completed, C. C.
(4) Command sample/hold to sample.

The sequence is repeated at a frequency which must be greater than twice the bandwidth limit of the analog signal.

There are several ways of practically implementing the above sequence of operations. Operational requirements such as the required analog signal bandwidth, the required digital data format (serial or parallel) and the nature of the subsequent digital processing govern the choice of conversion technique.

Successive approximation A/D converters are probably the most frequently used type in sampled data systems. They are capable of fast conversions and provide digital data in both parallel and serial form. Conversion time governs the time interval between samples, T_s and hence the sampling frequency, $f_s = 1/T_s$. The bandwidth of the analog signal must be limited to a frequency which is less than half the sampling frequency.

The conversion time, T_{conv}, of a n-bit successive approximation A/D converter is determined by the relationship

$$T_{conv} = (n+1) T_c$$

where T_c is the period of the clock pulses. The specific devices which are used as the functional elements in the system set the limit to the highest clock frequency that can be used. The fastest successive approximation converters currently available can make an 8-bit conversion in a conversion time of the order of 1 μs.

High speed successive approximation converters can be purchased as complete operational modules (for example, Burr Brown ADC60) or can be assembled at reduced cost using IC devices as the functional elements of the system. [6]

Tracking A/D converters are sometimes used in sampling and digitising an analog signal. Tracking converters give digital data in parallel form only. An advantage of a tracking converter is that it does not require the use of an analog sample/hold module in order to sample and digitise an analog signal. The tracking converter system itself performs the sample/hold function.

The digital output of a tracking converter is incremented continuously as it tracks or follows the value of its analog input signal. In order to take a digital sample value it is simply necessary to load an instantaneous value of the digital output word into an external digital storage resistor. The digital word can be held in the register for as long as it is required.

One of the main limitations of a tracking converter system when used in sampling and digitising analog signals is the slew rate limit of the system. The digital data produced by the converter follows the analog signal only provided that the rate of change of the analog signal does not exceed a limiting value. The maximum rate of change is

$$\text{slew rate limit, } SR = \frac{V_{fs}}{2^n} \times f_c \qquad (5.4)$$

Digital Processing of Analog Signals

where V_{fs} is the full scale analog range, n is the number of bits used and f_c is the clock frequency. Let us assume an 8-bit converter with V_{fs} = 10 V. The speed limitations of currently available devices limits the highest usable clock frequency to a value of the order of 10 MHz. This makes the maximum slew rate attainable with a tracking converter $10^7 \times 10/2^8$ V/s or $100/256 \approx 0.4$ V/s.

The slew rate limit of a tracking converter sets the maximum frequency of a full scale sine wave that can be accurately sampled and digitised as

$$f_{max} = \frac{SR}{2\pi V_{fs}/2} = \frac{f_c}{\pi 2^n} \qquad (5.5)$$

Slew rate limitations rather than sampling frequency limitations make the analog signal bandwidth of the tracking converter sampling system less than that which can be attained with a successive approximation converter system.

Another limitation of a tracking converter system is that while tracking the digital output dithers by 1 LSB either side of the 'correct' value. The quantisation noise present in the digital samples will therefore be greater than that obtained when using a successive approximation converter. The quantisation uncertainty is ± 1 LSB as compared with the ±$\frac{1}{2}$ LSB, of a successive aprroximation converter.

A third alternative in a sampled data system is to use an integrating-type A/D converter. High resolution integrating-type A/D converters are relatively inexpensive. An integrating converter produces a digital output which is a measure of the average value of the analog input signal during the conversion time; digital data is in parallel form. The conversion time of integrating converters is long when compared with that of the other techniques. Their use in sampling systems is thus limited to those systems in which analog signals are very slowly varying. Integrating converters do not of course require an input sample/hold module.

It should be remembered that whichever A/D conversion technique is used, the digital sample values will always have a quantisation error. Quantisation errors are reduced by using a converter with more bits. However more bits normally mean a slower conversion, hence a lower sampling frequency and a corresponding restriction of the bandwidth of the sampled analog signal. Quantisation errors are inherent—a practical sampling system may be expected to exhibit additional errors because of the non-ideal performance of the devices used in the system (see section 6.6).

A detail-by-detail comparison of the performance characteristics of presently available converter devices will not be given since it would rapidly become obsolescent. New converter devices are continually appearing on the market with the emphasis on an improved performance/cost ratio and greater user convenience. Readers are urged to consult manufacturers' product guides, application notes and device data sheets as the primary source of detailed (and often *free*) information about specific devices. Manufacturers are of course in the business of selling *their* products and can therefore be forgiven if they indulge in a certain amount of 'specmanship'—specifications are presented in the most favourable way, desirable performance characteristics are generally high-lighted and undesirable characteristics kept to the small print.

An example of a practical sampling and digitising system is given in the next section. It is intended to serve as a basis for the reader's own experimental evaluation. A practical familiarity with specific systems helps in acquiring the more general understanding of the techniques, which is necessary in order to make a critical appraisal of device data sheets.

5.4 SAMPLING AND DIGITISATION WITH A SUCCESSIVE APPROXIMATION A/D CONVERTER

An outline of a circuit for sampling and digitising an analog signal is given in figure 5.5. The system is an extension and modification of the successive approximation A/D converter described previously (figure 3.10). Sampling is performed by an inexpensive IC sample/hold which requires simply the external connection of a sampling capacitor. The choice of capacitor value is governed by the conflicting requirements of fast signal acquisition when switched to the sample mode and low rate of leakage when switched to the hold mode. [1, 5]

A monostable mutlivibrator and two NAND gates are used as a simple means of contrplling the sampling sequence. In applications not requiring the full 8-bit resolution, the system provides the option of a faster conversion by using a decreased number of conversion bits.

The system repetitively samples and digitises an analog signal which is applied to the input of the sample/hold. The $\overline{CC}$ signal produced by the continuously clocked successive approximation register initiates each sampling and digitisation sequence. Timing waveforms for 8-bit operation are shown in figure 5.6. the operation is as follows. $\overline{CC}$ goes low signifying the end of a conversion and at the same time triggering the dual monostable. The output of monostable 1, Q_1, goes high for a time T_1 and this puts the sample/hold in the sample mode. T_1, governed by component values C_1, R_1, is made a little longer than the acquisition time of the sample/hold. The pulse of duration T_2 generated by monostable T_2 is available to clock the register contents of the SAR into an external storage or shift register. The time period T_2 is made less than the period T_1. During the period T_1 the sample/hold is acquiring a new sample value; the SAR bits are not changing—they represent the digitally encoded value of the previously acquired sample. At the end of the period T_1, Q_1 goes low putting the sample/hold into the hold mode. $\overline{Q}_1$ goes high causing $\overline{S}$ to go low and the conversion of the held sample starts on the next low to high clock transition. The conversion is completed on the eight subsequent low to high clock transition.

The sampling time interval T_s is an exact number of the clock periods T_c

$$T_s = (8 + x) T_c$$

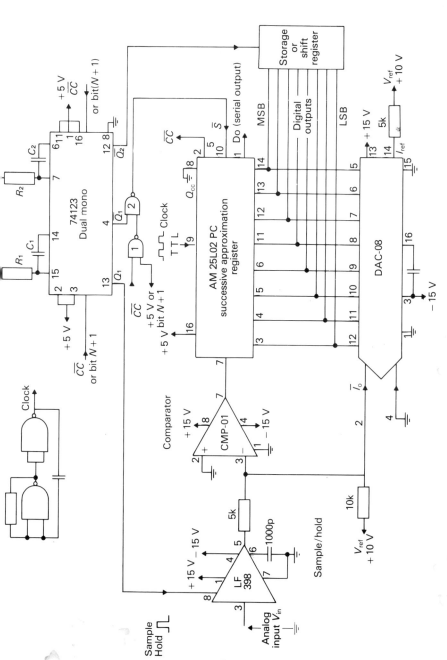

Figure 5.5 Sampling and digitising with a successive approximation A/D converter

where x is the smallest whole number such that $xT_c > T_1$. In the sequence shown in figure 5.6, $x = 3$, and $T_s = 11T_c$.

In reduced resolution applications with N bits ($N < 8$), the SAR is truncated by using the $N + 1$ register bit going low, instead of $\overline{CC}$, to initiate the sampling sequence. A possible stall up condition which could occur on first switching on is avoided by the inclusion of the NAND gate, 1. The gate ensures that $\bar{S}$ input to the SAR can be generated when either the $\overline{CC}$ or the SAR bit $N + 1$ are in the low state.

The comparator in figure 5.5 compares the DAC−08's $\bar{I}_o$ output with the input current. The $\bar{I}_o$ output current increments are turned on by a low logic level applied to the DAC's digital inputs. Digital output data from the conversion is thus encoded in a complementary offset binary code. This assumes that the usual convention of a high level use to represent logical 1 is employed. If the alternative convention of a logical 1 represented by a low level (active low) were to be adopted, the converter code would be interpreted as offset binary.

5.5 NON-LINEAR ENCODING

In a linear sampling and digitisation system the analog signal range is divided into 2^n evenly spaced quantised levels (n is the number of bits used in the digital encoding). All digitally encoded samples have the same quantisation uncertainty determined by the level separation. Quantisation uncertainty in a linear system is expressed as a percentage of the full scale analog range. If a linear system is to accurately digitise both large and small analog signals, very many closely spaced levels must be used (a large number of bits), otherwise the quantisation uncertainty may represent an unacceptably large error in the small amplitude signal samples. An alternative is to use non-linear encoding.

Linear and non-linear A/D conversion relationships are compared in figure 5.7; in each case the positive half of a 4-bit system is shown. In the non-linear system the quantised levels into which the analog range is dividing are closer together at the bottom of the analog range than they are at the top. The analog signal range is compressed: if a form of logarithmic compression is used quantisation uncertainty can be made to represent a fixed percentage of sample size rather than a fixed percentage of full scale as it is in the linear case. Non-linear encoding can be used to digitise both large and small sample values with the same modest accuracy.

Non-linear encoding can be performed in two ways. One way is to use analog signal compression followed by a linear A/D conversion. The other way is to use a non-linear A/D conversion in which a non-uniform quantisation of the analog range is introduced as an integral part of the encoding process.

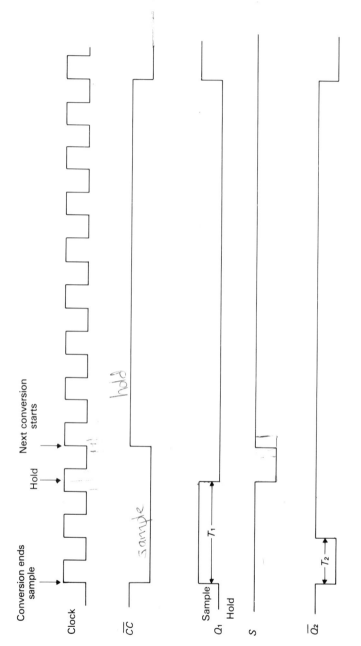

Figure 5.6 Sampling and conversion sequence for system shown in figure 5.5

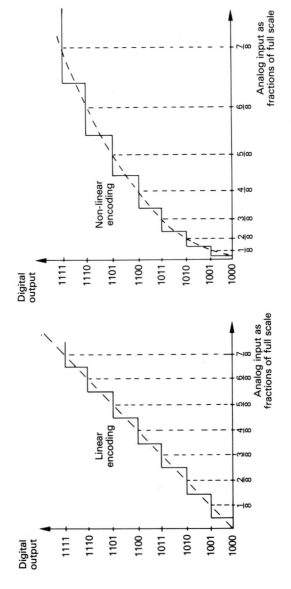

Figure 5.7 Linear and non-linear A/D conversion relationships; positive half of 4-digit system shown

5.5.1 Non-linear Encoding with a Log Amplifier

A functional schematic which illustrates the first method of non-linear encoding is shown in figure 5.8. A log amplifier is used to compress the analog signal range. Log amplifiers can be made with operational amplifiers [1] or they can be purchased as ready built functional modules (see Analog Devices or Burr Brown product guides).

The output of a log amplifier is related to its input by an equation of the form

$$V_o = -k \log_{10} \frac{V_{in}}{V_r}$$

In figure 5.8 we show a sign change and scaling adjustment. The output of the log amplifier which serves as the input to the linear A/D converter is

$$V_o' = k' \log_{10} \frac{V_{in}}{V_r}$$

The following points should be noted. Mathematically the log function is defined for positive numbers. A log amplifier will not accept a bipolar input signal. The output depends on the log of the ratio V_{in}/V_r where V_r is a reference value. The output of the log amplifier is zero when $V_{in} = V_r$, ($\log_{10} 1 = 0$); it is positive when $V_{in} > V_r$ and negative when $V_{in} < V_r$. The output responds to fractional changes in the input: every decade change in the input causes k_1 V change in the output. For example changes in the input of 0.01 V to 0.1 V, 0.1 V to 1 V and 1 V to 10 V all cause the same k V change in output. There is a lower limit to the input signal and zero input is unacceptable (as $x \to 0$ $\log_{10} x \to -\infty$).

Assuming the A/D converter in figure 5.8 is bipolar with offset binary code, the relationship between the output digital code $x_1, x_2, x_3, \ldots, x_n$ and the system input is

$$\left[(x_1 2^{-1} + x_2 2^{-2} + \ldots + x_n 2^{-n}) - \tfrac{1}{2}\right] V_{fs} = k' \log_{10} \frac{V_{in}}{V_r} \qquad (5.6)$$

where V_{fs} is the full scale analog range of the A/D converter. Knowing the range of V_{in} the system designer must choose scaling factor k' and reference value V_r to make the output range of the log amplifier coincide with the analog range of the converter.

The analog range of the converter is linearly quantised with level separation

$$\Delta V = \frac{V_{fs}}{2^n}$$

The corresponding separation of the input signal quantised levels is

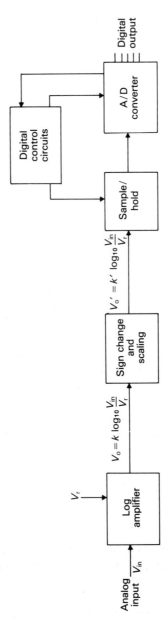

Figure 5.8 Log amplifier used in non-linear encoding

Digital Processing of Analog Signals

$$\Delta V_{in} = \frac{2.3}{k'} V_{in} \Delta V \tag{5.7}$$

The level separation increases with sample size. Not that $\Delta V_{in}/V_{in}$ remains constant. The logarithmic compression makes quantisation uncertainty a fixed percentage of the sample size rather than a fixed percentage of the full scale range.

A design example is now given. A log amplifier and a linear A/D converter are to be used to sample and logarithmically encode analog signals in the range 1 mV to 10 V. There is the requirement that the quantisation uncertainty shall never exceed 1 per cent of the sample value. The A/D converter has an analog range -10 to $+10$ V—it uses an offset binary code. We wish to know suitable values for the log amplifier scaling factor k' and reference value V_r. Also required is the number of bits needed to satisfy the accuracy specification.

For an input change 1 mV to 10 V (4 decades) the output of the log amplifier will change by $4k'$ V. This change is to coincide with the analogue range of the converter. We require

$$4k' = V_{fs} = 20 \text{ V}$$
$$k' = 5 \text{ V}$$

For the output range of the log amplifier to be centred at zero requires

$$V_r = 10^{-1} \text{ V}$$

This gives

$$V_o = 5 \log_{10} \frac{V_{in}}{10^{-1}}$$

that is $V_o = -10$ V when $V_{in} = 10^{-3}$ V $V_o = 0$ when $V_{in} = 10^{-1}$ V and $V_o = +10$ V when $V_{in} = 10$ V. We require

$$\frac{\Delta V_{in}}{V_{in}} \leq 0.01$$

that is

$$\frac{2.3}{k'} \frac{V_{fs}}{2^n} \leq 0.01$$

(from equation 5.7.)

$$2^n \geq \frac{2.3 \times 20}{5 \times 0.01} \quad \text{or } 2^n \geq 920$$

Note that $2^{10} = 1024$. Therefore a *10-bit* conversion is required. For the quantisation uncertainty in a 1 mV sample to be no greater than 1 per cent in a linear encoding system would require

$$\frac{\Delta V}{0.001} \leqslant 0.01$$

If an A/D converter with analog range of 0 to 10 V were used

$$\Delta V = \frac{10}{2^n}$$

This requires $2^n \geqslant 10^6$. Since $2^{20} \triangleq 1.05 \times 10^6$ a 20-bit linear encoding is needed!

It should be noted that with an equal number of bits a linear system gives greater accuracy than a non-linear system when the analog signal range is restricted to values near full scale. Non-linear encoding scores when wide range analog signals are to be encoded. A non-linear system gives the same modest accuracy throughout the signal range.

If logarithmic compression is used it must be remembered that the digital data is a logarithmic representation of the signal and it must be treated as such in any digital processing. In some systems the data is simply stored or transmitted and then returned to analog form. An antilog amplifier following an output D/A converter can be used to decompress the data and recover the original analog signal. [1]

Analog signal compression with a log amplifier is only suitable if input signals are unipolar. Bipolar signals require a compression characteristics which is symmetrical about zero. A characteristic is required which is logarithmic for the larger signal values but which passes linearly through zero. Analog functional relationships of this kind can be synthesised using operational amplifiers and complementary log transconductors. [17]

Precise analog methods for signal compression and decompression require a good deal of circuit trimming. It is difficult to match compression and decompression characteristics accurately. When bipolar signals are to be non-linearly encoded, a technique whereby the compression is introduced as an integral part of the encoding process is normally preferred. Such techniques require a DAC device having a defined non-linear conversion characteristic.

5.5.2 A Logarithmic DAC (Companding DAC) [22]

A monolithic logarithmic DAC manufactured by Precision Monolithics (the DAC–76) is a convenient device on which to base our discussion of non-linear decoding and encoding. A functional schematic for the DAC–76 is shown in figure 5.9. The device uses an external reference which may be varied. Considerations governing the reference amplifier and the way in which the reference voltage is applied are similar to those for the linear DAC–08 device described

Digital Processing of Analog Signals

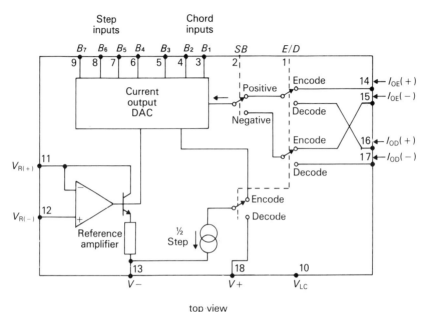

Figure 5.9 DAC-76 (Companding DAC) (Courtesy of Precision Monolithics)

in section 2.5. The digital inputs of the DAC-76 and the current outputs function in a different manner to those of the DAC-08.

The DAC-76 has two pairs of current output pins, $I_{OD\pm}$ and $I_{OE\pm}$, one pair is used in the decode mode of operation, the other in the encode mode. Selection between the decode and encode modes is performed by the logic level applied to the E/D input. A low logic level applied to E/D causes the output current to be steered towards the I_{OD} outputs, and a high level causes it to flow to the I_{OE} outputs.

The 8 data bits of the DAC-76 are used in a sign magnitude form of code. The sign bit determines which output pin the output current flows into. A high level applied to the sign bit causes the output current to flow into I_{OD+} of I_{OE+} dependent upon the E/D level. Output current flows into one or other of the I_{OD-} I_{OE-} pins when the sign bit is low. An additional $\frac{1}{2}$ step output current increment is switched in the encode mode. The reasons for this will be described shortly.

The basic connections which must be made to the DAC-76 in order to perform an expanding D/A conversion (decode operation) are shown in figure 5.10. An operational amplifier is used to convert output current into an output voltage. The polarity of the output voltage is determined by the sign bit. The operational amplifier acts as a current to voltage converter when the output current flows into the I_{OD+} pin. It acts as a unity gain voltage follower for a negative voltage applied to its non-inverting input terminal. Such a voltage arises as a result of the I_{OD-} current flowing through the resistor which connects the non-inverting input terminal of the amplifier to ground.

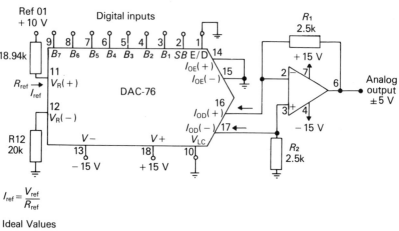

$$I_{ref} = \frac{V_{ref}}{R_{ref}}$$

Ideal Values
$I_{ref} = 528\ \mu A$
$I_{fs} = 2007.75\ \mu A$

	E/D	SB	B_1	B_2	B_3	B_4	B_5	B_6	B_7	E_0
Positive full scale	0	1	1	1	1	1	1	1	1	5.019 V
(+) Zero full scale 1 step	0	1	0	0	0	0	0	0	1	0.0012
(+) Zero full scale	0	1	0	0	0	0	0	0	0	0 V
(−) Zero full scale	0	0	0	0	0	0	0	0	0	0 V
(−) Zero full scale 1 step	0	0	0	0	0	0	0	0	1	−0.0012
Negative full scale	0	0	1	1	1	1	1	1	1	−5.019 V

Figure 5.10 DAC-76 (Companding DAC) (Courtesy of Precision Monolithics)

Digital Processing of Analog Signals 143

The idealised D/A conversion relationship (decode relationship) which is implemented is illustrated in figure 5.11. The complete decode transfer characteristic and, in enlarged scale, the characteristic for digital codes between 10000000 and 10100100 are shown. The sign bit simply fixes output polarity. The other 7-bits determine magnitude. Bits B_1, B_2, B_3 select between 8 binarily related chords or segments. Bits B_4, B_5, B_6, B_7 select one of the 16 linearly related steps within the chord selected by B_1, B_2, B_3.

A DAC which has a defined non-linear conversion characteristics can be used to implement a defined non-linear successive approximation A/D conversion (non-linear encoding). Readers are referred to the DAC–76 data sheet, [22] for details of the interconnections between a DAC–76, a comparator and a successive approximation register, which are needed in order to make a non-linear successive approximation converter.

5.6 APPLICATIONS OF SAMPLED AND DIGITISED ANALOG SIGNALS

The digital sample values obtained as a result of sampling and digitising an analog signal contain all the information which is present in the original analog signal. However, because the information is in digital form, it can be transmitted, processed and stored using precise digital circuit techniques rather than less precise analog techniques. The information can be reconstructed into analog form at any time using a DAC. Microprocessors provide very powerful and versatile digital signal processing but there are many applications possible without the use of microprocessor systems.

5.6.1 Digital Transmission of Audio Signals

Digital techniques for the transmission of audio signals have been in use for a number of years, for example, low noise military communication systems and the music distribution system to be found in some commercial aircraft. Digital sample values are transmitted serially. They are converted back into parallel form at the receiving end of the system before being applied to a DAC. Conversion of the audio signal into digital form and then back into analog form again may at first sight seem an unnecessary complication. The reason for using digital transmission is to gain noise immunity and to preserve the quality of the transmitted information. Analog signals tend to pick up noise and become progressively degraded when transmitted over long distances. Digital signals can always be restored by reshaping or regenerating pulses.

Some commerical aircraft have audio distribution systems which make any one of 8 different programmes available at each seat. It is of course possible to implement such a system using purely analog techniques. An analog system would require 8 analog signal channels wired in parallel to each seat. A digital system can be implemented using simply a two-wire system to each set. The digital system means simpler wiring, reduces weight and prevents possible crosstalk between channels.

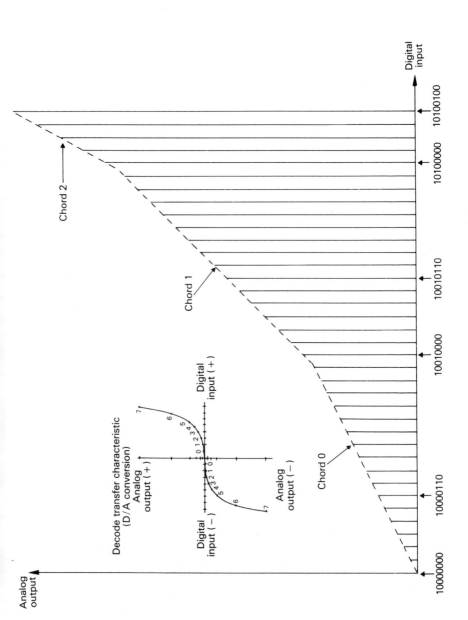

Figure 5.11 Decode transfer characteristic (D/A conversion)

Digital Processing of Analog Signals 145

An outline schematic of an aircraft digital audio distribution system is illustrated in figure 5.12. [17] Audio signals are normally stored on a multi-channel tape player—the channels are connected to an analog multiplexer. The multiplexer connects each analog channel in sequence to the sample/hold. The A/D converter digitises the samples and produces serial digital words corresponding to each of the 8 channels in sequence. A 3-bit address code is added to each digital word. The complete words are then sent down the two-wire system to every receiving unit in the aircraft.

Each receiving unit has a programme selector switch. The selector is linked to digital logic circuitry consisting of an address decoder and a serial to parallel converter. A DAC converts the parallel digital words into analog form. It converts only those words distinguished by their 3-bit address code corresponding to the selected programme.

Non-linear encoding and decoding provides a method of increasing the dynamic range of digital audio systems without increasing word length. [20, 21] The non-linear DAC–76 can be used in a two-way (send/receive) serial data transmission system. At the sending end of the system, a DAC–76 configured in the encode mode performs a non-linear A/D conversion (compressing encoding) of the sampled analog signal. At the receiving end, another DAC–76, configured in the decode mode, performs an expanding D/A conversion. Compression and expansion characteristics are equal and opposite thus giving a faithful reconstruction of the sampled analog signal. The logarithmic nature of the conversions maintains the signal to quantisation noise ratio appreciably constant for a wide range of analog signal amplitudes. [22]

5.6.2 Applications using a Semiconductor Shift Register Memory (Sequentially Accessed Memory)

If the sequence of digital sample values obtained as a result of sampling and digitising an analog signal are to be processed rather than simply transmitted, some method of storing the samples is required. A semiconductor sequentially accessed memory, made with static shift registers, provides a convenient and relatively inexpensive method of storgae. Shift register memories give a basis for many useful processing applications.

In order to fill a shift register memory with digital sample values the write inputs of the register are connected to theA/D converter's ‖ digital output lines. As each analog sample is taken, its digitally encoded value, when valid, is clocked into the register. An A/D converter shift register memory system can be used to produce a time delay in an analog signal. It can also be used to capture and store a transient analog event and is useful in such applications as digital waveform averaging, time compression by sampling, real time correlation and digital filtering. [17]

The functional schematic given in figure 5.13 illustrates the principles under-

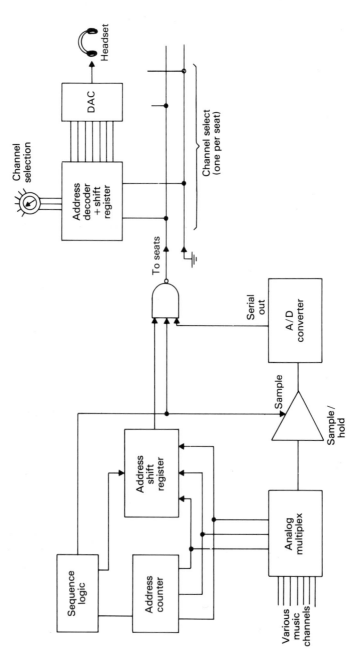

Figure 5.12 Aircraft music distribution system (Courtesy of Analog Devices)

Digital Processing of Analog Signals 147

lying the use of a shift register as a digital delay line, DDL. An analog signal is sampled and digitised by the sample/hold, A/D converter combination. The sampling frequency, in accordance with the sampling theorem, must be higher than twice the highest frequency component of the analog signal. Sample/hold acquisition time and A/D converter conversion time together set the limit to the highest usable sampling frequency. As the N-bit digital samples appear as parallel words at the A/D converter output they are clocked into the shift register and previously obtained samples are shifted along from left to right. K digital samples fill the register; sample values then appear at the shift register output and are reconstructed into analog form by the DAC. The reconstructed analog signal is a time-delayed version of the original analog input signal. The time delay T_d depends on the sampling frequency f_s and on the number of bits, K, in the shift registers, $T_d = K/f_s$ s.

An alternative to the use of parallel shift register is to use a single shift register into which digital sample values are loaded serially. Figure 5.14 shows a functional schematic for such an arrangement. The digital sample values are assumed to have N-bits. An N-bit serial in parallel out shift register is used at the end of the main delay register to convert the serial digital samples into parallel form. The parallel digital samples are applied to the output DAC via a latch. The latch holds the DAC digital input constant at the previous sample value while the next serial sample is clocked into the output shift register. The new sample value is clocked in parallel form into the latch and applied to the DAC. Using currently available shift registers and converters it is the clock frequency limit of the shift register that is likely to set the upper limit to the system's sampling frequency. The serial system is suitable for analog signals not requiring the highest possible sampling frequency.

DDLs find many applications in audio recording studios for creating a variety of audio effects obtained by playing back two identical signals at the same time with a time difference between them. Before the advent of electronic delay lines such effects were obtained by using two tape decks one of which had a variable speed. DDLs can be used to provide 'echo' or reverberation effects and also to multiply the sound of an individual instrument. For example, a violin sounded together with a time delayed version of say 5 ms sounds like two violins. Recording the composite sound and playing it back through a DDL yields a sound like four violins. It is possible to create an effect similar to an entire string section from one violin a DDL and a multi-track recorder.

The recirculate feature of a shift register memory allows other interesting applications of an A/D converter shift register memory system. It can serve as the basis for a transient analog event recorder and so provide a substitute for a storage type of oscilloscope. A transient capture and storage system is useful for studying the analog variables in many diverse research fields, for example biological signal analyses, mechanical destructive testing of components and shock wave analysis.

A functional schematic which illustrates the use of an A/D converter and

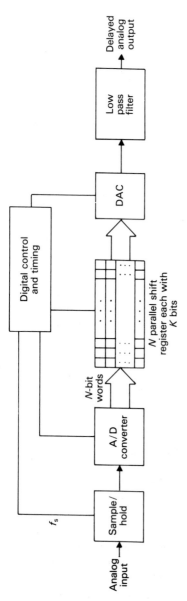
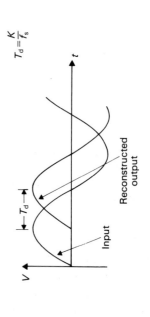

Figure 5.13 A shift memory used to produce a time delay in an analog signal

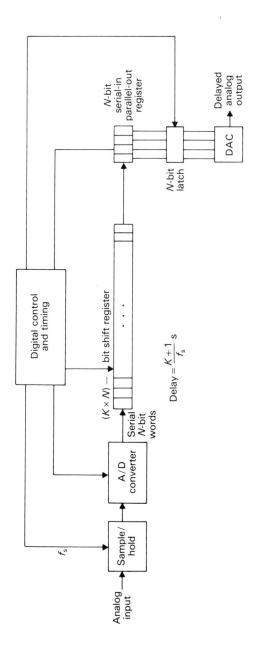

Figure 5.14 Shift register used as serial delay line

shift memory as a transient recorder is given in figure 5.15. The control and timing circuitry initiates a sampling sequence at the start of the transient signal which is to be recorded. The sampling sequence is stopped when the shift register is filled with digital sample values. The sampling frequency, f_s, is chosen so that a repetitive sampling of the analog signal occurs for the duration of the transient which is to be recorded. If the shift registers each have K bits the analog waveform which occurs in the time period K/f_s s is stored in the form of its digital sample values in the shift register memory.

A repetitive oscilloscope display of the stored transient is readily obtained. The shift register memory is simply switched to the recirculate mode. The stored data is then continuously recirculated and at the same time is reconstructed into analog form by the DAC to allow its oscilloscope presentation. Data can be recirculated using a shift register clock frequency quite different from the one used to clock the data into the memory. This allows the reconstruction and display of the analog transient to proceed at an arbitrary rate. A fast transient can for example be clocked out at a rate consistent with producing a permanent pen recorder display of the transient. Alternatively, if the original transient occurred slowly it could if required be clocked out at a faster rate to obtain a flicker-free oscilloscope display. If a permanent digital record of the transient is required the shift register contents are simply clocked into some form of permanent digital store.

The system shown in figure 5.15 can be used as the basis for generating quite arbitrary repetitive analog waveforms, for example, a shaped stimulus to be used in some analog process. The shift register is simply loaded with the required waveform. The waveform can be loaded in analog form via the A/D converter or, if required, the register can be loaded directly with digital information. The loading can be performed slowly (manually one sample at a time, should this be necessary). When the register is put in the recirculated mode the data can then be circulated continuously. It is reconstructed into a repetitive analog wave by the output DAC. The repetition frequency is controlled by the shift register clock frequency.

5.6.3 Using a RAM to Store Digital Samples

Random access memory devices, RAMs, provide an alternative to a sequentially accessed shift register memory as a means of storing the sequence of digital words obtained as a result of sampling and digitising an analog signal. RAMs can store a specific number of multi-bit words. It is usual to characterise the storage capacity of a RAM in terms of the number of words that can be stored in the memory and the number of bits in each word. A RAM designated as a 256 × 8 device has a storage capacity for 256 8-bit words. The total storage capacity is 2048 bits.

The physical nature of the storage elements in a RAM need not concern us—from the operational viewpoint it is convenient to regard data as being

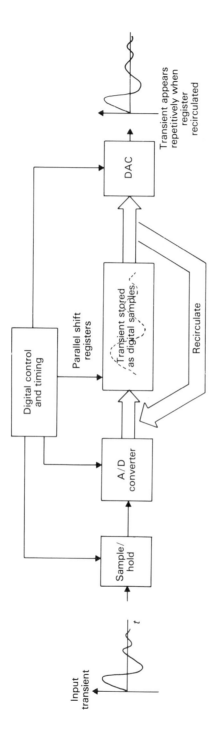

Figure 5.15 A/D converter and shift register used to capture and store an analog transient

simply stored at locations in the memory. A RAM allows digital data which is present at its input lines to be entered into a memory location—this is referred to as the *write* mode of operation. In the other mode of operation of a RAM, called the *read* mode, data stored at a memory location is caused to appear at the RAM's digital output lines. Selection of a particular memory location is accomplished by the application of a memory address word to the memory address decoder normally incorporated in the RAM. The decoder selects the memory location specified by the coded memory address word. In the read mode data stored at the specified location is routed to the RAM output lines. In the write mode data present at the RAM input lines is entered into the specified memory location.

Figure 5.16 is an outline schematic of an analog transient event recorder system which uses a RAM as the storage element. The RAM's memory address word is supplied by the parallel output of a binary counter. An m-bit counter is required, the 2^m different word memory locations of the RAM are addressed by the 2^m different output states of the counter.

In order to store and analog transient the RAM is put into its write mode. The transient, when it occurs, is sampled and digitised. Digital sample values are entered into the RAM memory locations. As each digital sample value becomes available it is written into a memory location. The memory address counter is then incremented up by one count selecting another location ready to store the next digital sample. This process is continued until all memory locations have been written into. The RAM then contains a permanent record, in the form of digital samples, of the analog transient occurring in the memory write period.

A DAC with its digital input lines connected to the RAM's output lines, can be used to reconstruct the analog transient from its stored samples. The RAM is put into its read mode of operation. The memory address counter is incremented up one clock pulse at a time causing the digital samples to appear in sequence at the RAM output lines. The DAC converts the samples back into analog form. As the memory address converter fills it recycles and the reconstructed analog transient is produced repetitively at the DAC output. The transient can be clocked out at a rate which is convenient for a particular type of analog display in use—a slow clock for pen recorder display, a faster clock for a repetitive oscilloscope display.

A fast RAM combined with a high speed parallel A/D converter (see section 3.11) can form the basis for a fast analog transient recorder. Systems of this kind can be used to observe the fast transient which occur say in shock and explosion testing. High speed converters combined with fast memories also find application in the digital processing of TV signals. [12]

5.7 CONVERTERS IN DATA ACQUISITION SYSTEMS

Electronic systems which are used to process analog signals and convert them into digital form are often referred to as data acquisition systems. The sampling

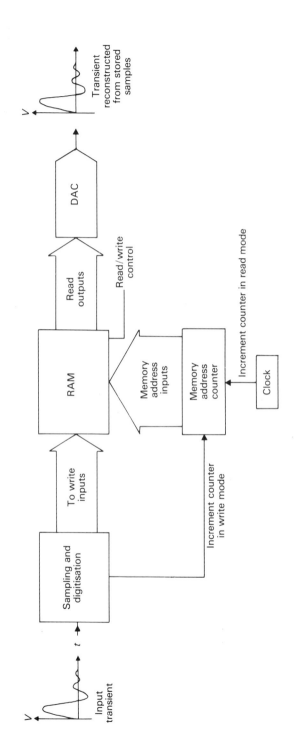

Figure 5.16 Transient recorder system using a RAM as the storage device

and digitisation systems discussed in previous sections are examples of single channel data acquisition systems. More generally electronic data acquisition systems are multi-channel systems. Multi-channel data acquisition systems perform the task of processing several analog input signals. They produce at the system's output the digital sample values corresponding to each analog input signal. The system's output is time shared between the various input channels in a manner which is controlled by the system.

The operation of switching several input signals one at a time to a single output channel is called *multiplexing*. A mechanical single pole multi-way switch can be regarded as a simple form of multiplexer device. However, in data acquisition systems the multiplexer operation is performed automatically under the control of digital logic signals. IC devices are available which perform the operation: there are analog multiplexer devices used to multiplex analog signals and digital multiplexers used to multiplex digital signals.

Figure 5.17 shows the interconnections between some of the functional components which might be included in a typical data acquisition system. The system is involved in acquiring measurement information. The physical parameters under investigation might for instance include the analog variables, pressure, temperature, strain or position. The first stage in each analog channel is a transducer which is used to convert the analog variable into electrical form. A separate amplifier is shown which is used to condition each transducer signal into a voltage in the range, say, 1 to 10 V prior to its sampling and digitisation.

Operational amplifier based circuitry is commonly used for conditioning transducer signals. The type of amplifier configuration required depends on the electrical nature of the transducer output signal. [1] In unfavourable electrical environments low level transducer signals can be obscured by unwanted interference signals and fully developed differential input measurements amplifiers may have to be used in order to reject common mode interference signals. [1, 9] There are a variety of other analog functional operations which it may be desirable to perform the analog signals prior to their sampling and digitisation. These operations are not shown in figure 5.17 but they could include squaring, linearising, multiplication by another analog variable and conversion to a RMS value. There are analog integrated circuits and modules available which can be used to perform such operations. [1, 9, 23, 24]

In figure 5.17 the conditioned analog signals are shown analog multiplexed as the input to a sample/hold module. The sample/hold and A/D converter are time shared between all the analog channels. An analog sample value, held by the sample/hold in the hold mode is converted into digital form by the A/D converter. During the conversion process the multiplexer can be instructed to select the next channel to be converted. When a conversion is completed the sample/hold is switched to the sample mode for a time which is long enough for it to acquire the next channel value. It is then returned to the hold mode and the sequence is repeated.

The multiplexer may be instructed to select analog channels in sequence

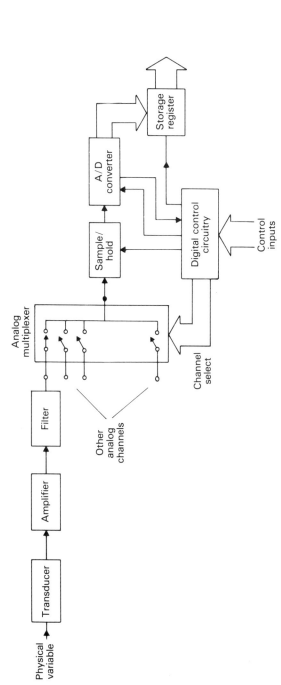

Figure 5.17 Multi-channel data acquisition system

or in a random access mode. In systems in which some of the analog variables are changing more rapidly than others it may be desirable to sample those channels more frequently than the others. Selection of a particular input channel is performed by a binary coded digital word which is applied to the multiplexer by the digital control circuitry. The number of bits, N, required in the channel select word depends on the number of multiplexer channels (number of channels $= 2^N$). An 8-input multiplexer needs a 3-bit channel select word.

Figure 5.17 shows just one of the many possible arrangements of the functional elements of a multi-channel data acquisition system. Arrangements differ in respect of the number if the system elements which are shared by all the analog channels. The extent of the sharing is determined by the stage in the system at which the multiplexing operation is performed. Maximum sharing is obtained in systems in which low level analog multiplexing is performed right at the input end of the system. In such systems all system elements, apart from transducers, are time shared between the analog channels. At the other extreme, in a system in which there is no sharing each channel has its separate analog conditioning circuitry, its own sample/hold and its own A/D converter. In such a system the multiplexing operation is performed on the signals when they are in digital form. There is clearly a range of possible configurations between these two extremes.

Systems in which transducer signals require amplification by a differential input instrumentation amplifier can share a single instrumentation amplifier between channels. A differential analog multiplexer is used to connect each differential signal, in turn, to the amplifier differential input terminals. Details of examples of such systems are to be found in manufacturers' data sheets and application notes, for example, National Semiconductors data sheet for type LF11508 8-channel analog multiplex; Burr Brown data sheet for data acquisition system type MP20 (a self-contained data acquisition system suitable for microprocessor interfacing).

In some experimental situations the parameters which characterise the state of the experimental system must all be measured at the same instant in time. Coincident measurements require the use of a data acquisition system with a sample/hold in each analog channel; an outline system configuration is shown in figure 5.18. All sample/holds are simultaneously switched to the hold mode. They remain in the hold mode while their outputs are analog multiplexed to the A/D converter in the desired sequence. When each held analog sample value has been digitised the sample/holds may be instructed to acquire new sample values. Clearly the sample/holds used must be capable of accurately holding for the time taken by the A/D converter to perform the several conversions—a more demanding requirement of the sample/hold than when it is only required to hold for a single conversion.

The traditional approach to configuring a data acquisition system has been directed towards maximising the number of shared elements. This approach had obvious economic advantages when data acquisition components, parti-

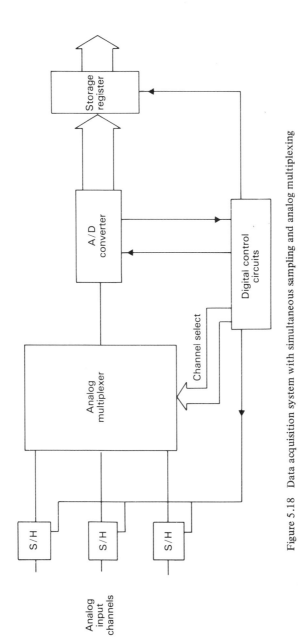

Figure 5.18 Data acquisition system with simultaneous sampling and analog multiplexing

cularly A/D converters, were very expensive. The availability of inexpensive integrated circuit converters, amplifiers and sample/holds lessens the economic advantages of the traditional approach. It makes data acquisition system configurations of the form outlined in figure 5.19 economically viable. The use of separate analog conditioning circuitry and a separate A/D converter per channel may at first sight appear extravagant but nowadays it may turn out to be the most cost-effective approach.

For a given system throughput (throughput is the number of samples per second which can be handled) in the converter per channel approach the converters can be slower and therefore less expensive than the single converter used in an analog multiplexed system. The converter per channel approach may make it possible to dispense with sample/holds when signals are slowly varying.

The successful transmission and multiplexing of low level analog signals can involve a great deal of engineering effort. The converter per channel approach and digital multiplexing, although it uses more components, is easier to implement. In data acquisition systems in which transducers are widely separated and remote from the data collection centre the analog signals can be digitised right at their source. Data can then be transmitted digitally, in serial form if required, with all the advantages of increased noise immunity when compared with analog transmission.

The choice of a suitable data acquisition system configuration for use in a specific application, like many other electronic design decisions must start with as full a characterisation as possible of the task which the system to perform. This must be followed by a consideration of the design options which are available to fulfil the desired system performance specification. It may be that one of the ready assembled, single circuit board data acquisition systems which are available will be found suitable. The designer will need to balance the savings in design and engineering time obtained by using a ready assembled system against the increased inital cost when compared with component level system assembly. A detailed study of manufacturer's up-to-date data component product guides is essential. New data conversion components and ready assembled systems are continuously appearing on the market. IC converter products are taking over tasks formerly performed by more expensive circuit modules. Data acquisition system design approaches which were once very expensive may, when implemented with monolithics, turn out to be the most cost-effective designs. Some of the factors which require consideration when specifying the performance characteristics of a data acquisition system are discussed in the following sections.

5.8 DATA ACQUISITION SYSTEM CHARACTERISATION

The basic parameters which characterise a data acquisition system are

(1) the number of analog channels

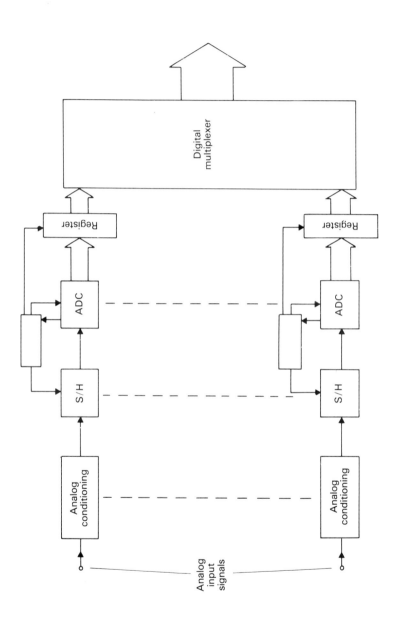

Figure 5.19 Data acquisition system configuration using one converter per channel

(2) throughput and analog signal bandwidth (speed)
(3) accuracy.

System throughput is the rate at which the system can generate digital sample values—it governs the rate at which the analog signals can be sampled (the sampling frequency). According to the sampling theorem a channel sampling frequency must be greater than twice the highest significant frequency component of the analog signal which is applied to that channel. This minimum sampling frequency is necessary if the sampled signal is to contain all the information required for its reconstruction.

Sampling frequency limitations are not the only factor governing analog signal bandwidth. Data acquisition systems have an input slew rate limitation. The slew rate of an analog signal is its rate of change. In a data acquisition system the rate of change of the analog input signals must not exceed a limiting value otherwise digital sample values are in error. Slew rate limitations can make the analog signal bandwidth for full scale analog signals considerably less than the bandwidth allowed from sampling theorem considerations.

A data acquisition accuracy specification gives the accuracy with which digital sample values represent the instantaneous analog sample values of the input signal. System accuracy is expressed for the system operating at its specified throughput. Conventionally, accuracy specifications are quoted in terms of percentage error; strictly speaking they represent inaccuracy specifications.

In all data acquisition systems there is invariably a speed/accuracy design compromise—increasing the one means trading off the other. Often another overiding constraint is cost. The cost of a data acquisition system is related to both its speed and accuracy—accurate, high speed data acquisition systems are expensive. It is therefore uneconomic to specify a data acquisition system with greater speed or accuracy than is strictly necessary.

5.8.1 Accuracy Considerations

The functional elements in a data acquisition system are not ideal. The practical performance limits and errors of all the elements of the system influence the overall performance and accuracy of the system. A/D converters are usually the most expensive elements in a data acquisition system; other functional elements should normally be chosen to have a higher accuracy than that of the A/D converter. This will ensure that overall accuracy is not significantly degraded below that of the most expensive system element.

A/D converters, even ideal ones, have an inherent $\pm \frac{1}{2}$ LSB quantisation error which limits their resolution and accuracy. There are additonal errors associated with any practical implementation of the A/D conversion process. A/D converter data sheets seldom give an overall accuracy specification—they list performance errors separately. Some of these performance errors are given below. (Readers requiring a further discussion of performance characteristics are referred to section 6.6)

Scale error

The constant of proportionality between analog input and digital output may be different from its design value because of errors in the reference voltage, resistor values and amplifier gain.

Offset error (zero error)

The analog input for a zero digital code output may be other than zero. Offset arise because of amplifier and comparator input offset voltage and bias current.

Scale errors and offset errors can normally be trimmed out at a particular temperature but errors due to their temperature dependence remain.

Non-linearity

Practical A/D converters exhibit departures from the ideal linear analog input/ digital output transfer curve. Linearity is a performance characteristic which is intrinsic to a specific converter and it cannot be externally adjusted. It sets the attainable accuracy limit of a converter.

It should be noted that quoted performance errors are in addition to the inherent $\pm \frac{1}{2}$ LSB quantisation error. A measure of the best accuracy that can be obtained with a properly adjusted converter is obtained by adding the non-linearity error to the inherent $\pm \frac{1}{2}$ LSB quantisation error. Performance errors are normally expressed as a fraction of the LSB or as a percentage of full scale. An 8-bit A/D converter stated to have an overall conversion accuracy of ± 1 LSB would have an error $\pm (1/2^8) \times 100$ per cent of full scale (± 0.39 per cent full scale). That is twice the minimum possible quantisation error of $\pm \frac{1}{2}$ $V_{fs}/2^8 = \pm 0.195$ per cent of full scale.

The functional elements used for analog signal conditioning (amplification), prior to conversion, may be expected to introduce scaling and offset errors. Such errors arise as a result of resistor tolerance and there are also gain and offset errors associated with operational amplifier based conditioning circuitry. [1] If an analog multiplexer is used there is a possible signal loading error due to the finite ON resistance of the multiplexer switches. Switch ON resistance forms a potential divider with the input resistance of the following system element. Sample/holds have gain and offset errors. There are additional time-dependent errors which depend on sampling frequency (see section 5.8.2).

It is not too difficult to achieve errors in analog conditioning circuitry and sample/holds which are less than an 8-bit converter's LSB. The LSB magnitude for a 12-bit converter is $(1/2^{12})V_{fs}$, that is 0.024 per cent of full scale and the performance requirements of the analog conditioning circuitry are more demanding. Clearly the accuracy requirements for analog conditioning prior to a 16-bit A/D conversion are very stringent indeed. However, if the accuracy capabilities of high resolution converters are to be achieved it is essential that errors introduced by other elements in the system be less than the conversion error.

Data Converters

5.8.2 Speed Considerations

A finite time is required to perform the various functions involved in the operation of a data acquisition system. When operations are performed sequentially the time delays associated with each operation add to give the total time required to obtain each digital sample. The rate at which digital samples are generated is determined as the inverse of this time for one sample.

Successive approximation converter with no sample/hold

An A/D converter takes a finite time to perform a conversion: its conversion time T_{conv}. If connected so as to perform continuous conversions it produces $1/T_{conv}$ digital samples/s. If the converter is time shared between N analog channels with an analog multiplexer connecting each channel in turn to the converter, there is an additional time delay associated with obtaining each digital sample. This is the time taken by the multiplexer to switch channels plus the time taken by the multiplexer output to settle following a possible transient output switching error, T_{mux}.

In such a system, using no sample/holds the maximum throughput rate per channel assuming all channels are sampled at the same rate is

$$\text{max throughput rate (with no sample/hold)} = \frac{1}{N(T_{conv} + T_{mux})} \text{ samples s}^{-1} \text{ channel}^{-1} \quad (5.8)$$

An A/D converter with $T_{conv} = 10\ \mu s$ used with an 8-channel ($N = 8$) analog multiplexer with $T_{mux} = 3\ \mu s$ would give

$$\text{max throughput rate} = \frac{1}{8(10+3) \times 10^{-6}} = 9.6k \text{ samples s}^{-1} \text{ channel}^{-1}$$

Sampling theorem considerations allow an analog signal bandwidth just less than half this throughput rate. However, input slew rate limitations make the allowable analog signal bandwidth for full scale signal much less than this figure.

Digital sample values are required to represent the analog signal value at the instant the converter is instructed to start conversion. An accurate representation demands that any change in the analog signal which takes place during the conversion time be no greater than the converter's LSB analog magnitude. Accuracy requirements thus limit the maximum rate of change of the analog signal to a value

$$\left| \frac{\Delta V_{in}}{\Delta t_{max}} \right| = \frac{V_{fs}}{2^n T_{conv}} \quad \text{V/s} \quad (5.9)$$

for a successive approximation converter when it is not preceded by a sample/hold.

The slew rate limits the allowable bandwidth of analog input signals. Consider a full scale sinusoidal input signal

Digital Processing of Analog Signals

$$V_{in} = \frac{V_{fs}}{2} \sin 2\pi f t$$

Its maximum rate of change (obtained by differentiation) is

$$\left|\frac{\Delta V_{in}}{\Delta t}\right|_{max} = 2\pi f \frac{V_{fs}}{2}$$

Thus if conversion accuracy is not to be degraded by slew rate induced error, a full scale sinusoidal signal must have a frequency no greater than f_{max} such that

$$2\pi f_{max} \frac{V_{fs}}{2} = \frac{V_{fs}}{2^n T_{conv}}$$

That is

$$f_{max \atop (fullscale)} = \frac{1}{\pi 2^n T_{conv}} \qquad (5.10)$$

For an 8-bit A/D converter with $T_{conv} = 10\ \mu s$ this gives

$$f_{max} = \frac{1}{\pi \times 2^8 \times 10^{-5}} = 124\ Hz$$

Quite a severe limitation to full scale analog signal bandwidth! The maximum frequency for a sinusoidal signal with peak-to-peak size $V_{pp} < V_{fs}$, if accuracy is not to be degraded by slew rate induced error, is

$$f_{max} = \frac{V_{fs}}{V_{pp} \pi 2^n T_{conv}} \qquad (5.11)$$

Sample/hold Increases Speed

Input slew rate limitations are less restrictive when an A/D converter is preceded by a sample/hold. Ideally the value of the analog input signal which exists at the instant that the hold command is applied is held constant while it is converted. In a practical sample/hold there is a time delay between the application of the hold command and the time at which the output of the sample/hold is no longer affected by changes in the input signal. This time delay is called the *aperture time*, T_{ap}, of the sample/hold. It arises because of the characteristics of the solid state switches which are used in sample/holds. The hold command causes the switch to go from its low resistance ON state to its very high resistance OFF state. Resistance change does not take place instantaneously and while switch resistance is increasing changes in the sample/hold input signal may still affect its output.

Conversion accuracy requires that any change in the input signal which occurs during the sample/hold aperture time should be no greater than the converter's LSB analog magnitude. This makes

$$\left| \frac{\Delta V_{in}}{\Delta t} \right|_{max} = \frac{V_{fs}}{2^n T_{ap}} \qquad (5.12)$$

(with sample hold)

where T_{ap} = sample/hold aperture time, n = number of bits in converter code and V_{fs} = converter full scale input range. Comparison with equation 5.9 — the expression for input slew rate for a system which does not use a sample/hold — shows a speed improvement in proportion T_{conv}/T_{ap}. Available sample/holds may be expected to have $T_{ap} \ll T_{conv}$.

In a data acquisition system in which analog signals are analog multiplexed into a shared sample/hold and A/D converter, (see figure 5.17), the channel throughput rate is

$$\text{Max throughput rate} = \frac{1}{N(T_{conv} + T_{ap} + T_{aq})} \text{ samples s}^{-1} \text{ channel}^{-1} \quad (5.13)$$

Where N is the number of channels (assumed all sampled at the same rate) and T_{aq} is the acquisition time of the sample/hold. Sample/hold *acquisition time* is the time taken for the output signal of the sample/hold to become equal to its input signal (within a specified error) when the sample/hold is switched from the hold to sample mode. Note that the multiplexer switching time is not included in the above expression for throughput rate. It is assumed that $T_{mux} < T_{ap} + T_{conv}$ and that the multiplexer is instructed to select the next channel while the converter is performing a conversion on the previously selected and held sample. Accuracy requirements dictate that any droop in the sample/hold output which occurs during the conversion time should be no greater than the converter's LSB analog magnitude $V_{fs}/2^n$.

It is obvious that using a specific sample/hold and A/D converter (fixed T_{aq} and T_{conv}) throughput rate per channel varies inversely with the number of shared channels. The greatest possible throughput per channel is obtained in a system in which there is no sharing — in a system in which there is a separate sample/hold and A/D converter in each channel with digital multiplexing of the converter outputs.

5.9 INTERFACING CONVERTERS TO MICROPROCESSORS

The present upsurge in the importance of data converters stems directly from the ready availability of low cost microprocessor/microcomputer systems. The majority of the new areas of application for data converters are directly concerned with getting information into or out of a microcomputer system. The idea of using a digital computer to measure, monitor and control processes or events in the outside world is by no means new. However, the areas in which the idea has actually been applied have formerly been limited by economic considerations (high cost of large computer systems and data converters).

Low cost microprocessors and converters now make it economically viable to provide automatic operation of small scale measurement and control systems and even of individual instruments. Microprocessor controlled systems require a minimum of attention by human operators — the microprocessor provides the 'intelligence' for these systems. Intelligent microprocessor controlled electronic systems have been developed and are now being developed for use in many diverse measurement and control applications, for example, industrial control, environmental control, weather monitoring, medical electronics, car control systems and perhaps control systems for the home of the future. [25]

Analog signals are involved in some part of most measurement and control systems. Transducer signals and the signals which are required to activate control elements are usually analog signals. Human operators can appreciate information trends more rapidly when the information is presented in analog fashion. A/D converters are needed to get transducer signals into a microprocessor. A microprocessor needs D/A converters to generate analog signals to actuate analog control elements or to provide an analog display of information.

In principle any A/D or D/A converter device can be connected to a microprocessor to provide an analog digital interface. In practice the ease with which a specific converter device can be mated to a microprocessor (that is, the cost and complexity of the interconnection circuitry) differs widely between different converter devices.

At the time of writing many of the available converter devices are not readily interfaced to a microprocessor. The additional hardware which they require to effect a transfer of data between mircoprocessor and converter can cost as much as the data converter itself. The reasons for this lack of compatibility largely arise from the separate development of microprocessors and converters. A microprocessor, in a gross oversimplification, is a digital computer that has been shrunk down to a 'micro'/miniature size. Microprocessors stem from developments in very large scale integration, VLSI, in the digital computer field — mainly the province of the large multi-national semiconductor firms. Historically converters have been developed by smaller manufacturers who have specialised in data acquisition products.

All manufacturers have now recognised the present and future potentially large size of the market for microprocessor, compatible data converters and microprocessor controllers data acquisition and distribution systems. They are rushing in to meet the demand and get a slice of the market. However the reader should examine carefully a manufacturer's claim that a particular converter device is microprocessor compatible. It can mean that the device can be simply hooked up to the microprocessor's data bus and control lines but often it is not quite that simple and can involve the use of several interface chips — there is complete compatibility, marginal compatibility and downright incompatability (it's rather like marriage!).

The problem with interfacing converters and microprocessors is that there is no such thing as a standard microprocessor. The currently available microprocessor systems which have already received wide acceptance differ substantially in the way in which they manipulate data and control its transfer between the various system elements. The question which the designer must ask himself when looking for a converter to interface with a microprocessor is, 'How compatible is this particular converter device with the microprocessor which I am using or propose to use?' An answer to the above question can only be sensibly formulated if the designer has a fairly detailed knowledge and understanding of the workings of the particular microprocessor system.

A treatment of microprocessor system architecture and operation is clearly outside the scope of this book. The designer must look to microprocessor manufacturers' instruction manuals and application notes for detailed information about specific microprocessor systems. In the discussion which now follows, we treat some of the general considerations involved in a microprocessor-converter interface. Examples of specific microprocessor compatible converter devices are mentioned to tie the discussion down and to give the reader some idea of the range of devices which are available at the time of writing. A basic familiarity with microprocessor terminology is assumed.

5.9.1 Interfacing Considerations

A data converter-microprocessor interface involves both hardware and software: hardware in the form of interconnect circuitry and perhaps involving additional IC packages to establish compatibility between data converter and microprocessor signals; software in the form of the instructions which must be written into the system's programme to make it accommodate the data transfer as part of its normal operating procedure.

Three types of digital signal may be distinguished in a microprocessor system: data, address and control signals. The signals are exchanged between the system's functional elements via groups of lines referred to respectively as the data bus, the address bus and the control bus.

The data bus handles the digital signals which represent the information which the microprocessor system is operating on. The address bus carries digitally coded signals which signify the location in the system of the particular data word on which the system must operate at that particular time in its operation cycle. Control signals are used to initiate system operations.

A data converter-microprocessor interface deals with all three types of signal. Transfer of data between converter and microprocessor system is accomplished via the microprocessor data bus. Address signals are required to signify that a particular converter, as distinct from another converter or another system

Digital Processing of Analog Signals 167

element, is required to exchange data with the bus. Control signals are required to initiate the data exchange. In the case of an A/D converter–microprocessor interface, there is one control signal required to instruct the converter to perform a conversion and another control signal is required to indicate when the conversion has been completed and effect data transfer to the data bus.

In a microprocessor system the logic gates which are connected to the data bus, in addition to their normal high and low logic states, have a high impedance state that acts like an open circuit. This feature allows a three-state data bus to be shared by all the functional elements in the system. Functional elements not involved in data exchange with the bus at any time are put in their high impedance state, in effect temporarily disconnecting them from the data bus. Exchange of data between a microprocessor data bus and a converter has to be similarly effected by way of data registers or latches which are compatible with the three-state data bus. Such buffer registers in effect isolate the converter from the data bus at all times other than those at which data exchange is instructed to take place. A data converter which is classed as microprocessor compatible may be expected to include a buffer register for connection to the microprocessor data bus.

Another aspect of data bus compatibility is the width of the bus. Microprocessor data buses consist of 4, 8 or 16 bits. If the data bus has more bits than the converter words, unused data bus lines can in effect be ignored. However, if the converter word is wider than the microprocessor bus, some means of in effect multiplexing the data between the converter and the bus must be used.

The way in which Analog Devices 12-bit A/D converter model AD574 achieves compatibility with an 8-bit microprocessor bus is illustrated in figure 5.20. The

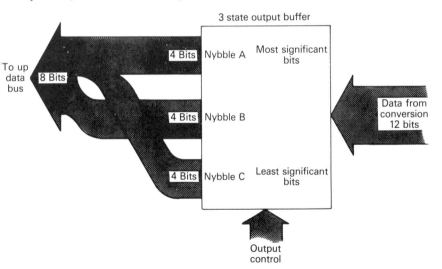

Figure 5.20 12-bit A/D converter interfaces to 8-bit microprocessor data bus. Output buffer used in Analog Devices converter model AD574

device incorporates a three-state output register which can be operated in such a way that it can multiplex data directly to an 8-bit bus. The register's 12-bit output word is divided into three smaller 4-bit words referred to as 'nybbles' (small 'bytes'). The digital lines of output nybble A which contain the 4 highest significant bits are paralleled with nybble C containing the 4 lowest order bits. Control circuitry simultaneously enables nybbles A and B and disables C thus putting the 8 higher order bits obtained in a conversion on the data bus. Nybble A is then disabled, nybble C enabled and 4 '0's made to appear at nybble B. This now puts the 4 lower order bits and 4 trailing '0's on the data bus.

An increasing number of DACs which incorporate an input data latch are becoming available. Examples of 8-bit devices are Motorola MC6890 and Analog Devices AD7524. The external connections required to implement an AD7524 application are illustrated in figure 5.21. The device may be hooked directly to

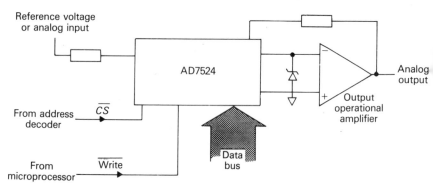

Figure 5.21 DAC with input data latch interfaces directly to microprocessor data bus: Analog Devices DAC model AD7524

an 8-bit microprocessor data bus. It appears to the microprocessor as a memory location into which data on the bus is written in a microprocessor write cycle. Control signal requirements consist of active low 'write' and 'chip select'. In most microprocessor applications, no additional circuitry apart from the normal chip select address decoder is required. The AD7524 requires an external reference and operational amplifier. More recently Analog Devices have introduced the AD588 'DACPORT', an 8-bit microprocessor compatible DAC which in addition to an input data latch incorporates its own internal output operational amplifier and reference.

National Semiconductors have introduced a series of multiplying DACs (MicroDAC family) which provide a double buffering of digital input data. [26] The input buffering and control signal arrangements used in these devices are illustrated in figure 5.22. The converter's register holds constant any data undergoing conversion while the input latch acquires new data. A feature of the double buffering arrangement is that it allows a 10-bit converter input to be assembled from two microprocessor data bytes. The devices use an external reference and

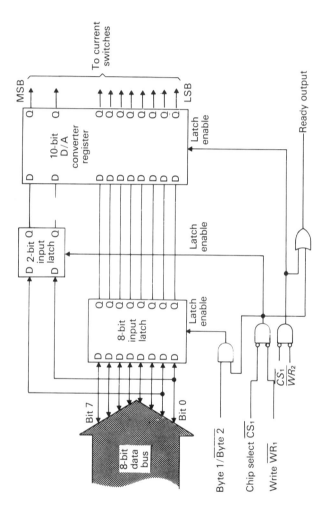

Figure 5.22 Input buffering arrangement used in National Semiconductors Micro DAC family of multiplying DACs

external output operational amplifier. Controlled gain and attenuation applications (see section 4.2) in which a microprocessor effects the control can easily be arranged. [26]

Most of the currently available microprocessor compatible converter devices are designed to operate in such a way that exchange of data between converter and microprocessor system is effected by the microprocessor treating the converter as a memory element. The technique is referred to as *memory mapped* I/O (input/output). There are other more specialised techniques but they require either specialised instructions or a considerable amount of extra hardware. [27, 28] Memory mapped converters keep programming simple and make for application flexibility.

Analog Devices A/D converter model AD7574 is an example of a low cost device designed specifically for ease of interface with microprocessors. A functional schematic for the AD 7574 is given in figure 5.23. It is a successive approx-

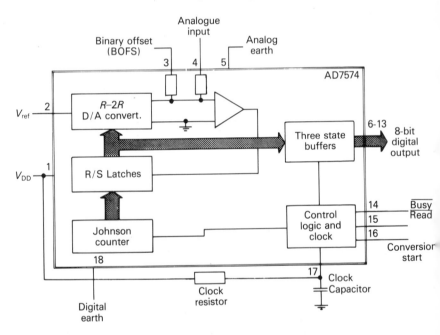

Figure 5.23 Functional schematic for Analog Devices AD7574 low cost microprocessor compatible 8-bit A/D converter

imation converter which uses a Johnson counter and R-S latches to implement the successive approximation logic. It can complete an 8-bit conversion in 15 μs. The device allows three different operating modes determined by the connections which are made between the converter's conversion start, read and busy control lines and the microprocessor system.

The AD7574's operating modes are referred to as the **RAM** mode, the **ROM**

Digital Processing of Analog Signals

mode and the slow-memory mode. In the RAM mode the converter appears to the microprocessor as a location in a random access memory that must be written into before it can be read. The microprocessor memory write command is used to start a conversion and its read command is used to transfer the data obtained as a result of the conversion. In the ROM mode the converter acts as a location in read only memory. A read command issued to the converter by the microprocessor puts the results of a preceding conversion on to the data bus, resets the converter and initiates a new conversion. Figure 5.24 shows an AD7574

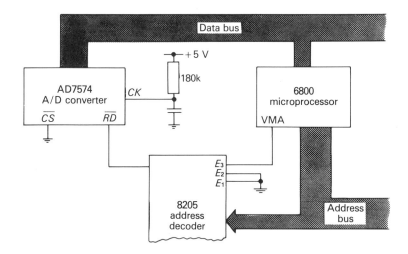

Figure 5.24 AD7574 interfaced to 6800 microprocessor using ROM mode of operation

interfaced to a 6800 microprocessor with connections appropriate for the ROM mode of operation. In the slow-memory mode of operation the converter begins a conversion when the microprocessor addresses its memory location. At the same time it sends a busy signal to the microprocessor indicating that it must wait until data is available for reading.

Several AD7574s can be interfaced to a microprocessor. Each converter is assigned its own particular address code. When a converter address code appears on the microprocessor address bus and the converter receives the control signals appropriate to its mode of operation, it puts data obtained as a result of a conversion on to the microprocessor data bus. Readers requiring detailed information about the AD7574 and the limitations and advantages of its various modes of operation are referred to manufacturers' literature on the device.

In the examples of data converter–microprocessor interface that have been given so far, the digital data has been transmitted between microprocessor and converter in parallel form. There is of course the alternative of serial transmission of data between converter and microprocessor. Serial transmission via a single pair of twisted wires becomes an attractive proposition when the point at which data is acquired is a long way from the microprocessor. However if the highest possible sampling rates are required a parallel interface is necessary and the converter should be located close to the microprocessor.

In most microprocessor families there are devices available which are specifically designed to handle data interfacing. So-called programmable peripheral interface devices (or something similar) provide an alternative means of handling parallel data transfer. Serial data transfer to a microprocessor system may be accomplished using a device called a *universal asynchronous receiver transmitter*, UART, which will accept parallel data and transmit it serially or will receive serial data and put it into parallel form. A serial interface between a converter and a microprocessor generally involves the use of a UART at both sending and receiving ends. UARTs interface readily with microprocessors but the interface between a converter and a UART often needs extra hardware. Detailed considerations involving UART operation are best obtained from manufacturers' data sheets and application notes.

The discussion that has been presented has centred around the interfacing of individual converters to a microprocessor. Also available are ready built multichannel data acquisition and data distribution systems which are designed for easy microprocessor interfacing. Such systems are appearing as single packet devices (for example Analog Devices AD7581). The general trend in mircoprocessor orientated data converter products is for manufacturers to provide increasingly complex single packet subsystems requiring very few external components. As subsystems grow in complexity the user may increasingly concern himself with complete system problems, leaving the problems of subsystem design to the device manufacturers.

Some microprocessor manufacturers have already gone as far as putting data converters on the same chip as a microprocessor. An example of such a device is the Intel model 2920, a so-called analog microprocessor. [30] At the present time, devices such as this are directed towards somewhat specialised applications. However, with the rate of development of microprocessor compatible data converter products being what it is, there is no telling what may become available even in the time it takes this book to get into print. The only really up to date sources of information about which data converters are most easily interfaced to a microprocessor are the manufacturers themselves and review articles in current trade journals. Some useful review articles, available at the time of writing, are included in the list of references. [28, 31, 32]

5.10 SELF-ASSESSMENT EXERCISES

1. Fill in the blanks in the following statements.
   (a) A multiplexer is a device which is used to ....
       to a single output channel.
   (b) Data acquisition system is the name given to an electronic instrumentation system which is used to ....

Digital Processing of Analog Signals 173

(c) In an ideal sampling and digitisation system there is a possible error in digital sample values because of ....
(d) A simple model of a sample/hold circuit consists of ....
(e) The output of a sample/hold when it is in the sample mode is ideally
(f) In the hold mode the output of a sample/hold is ideally ....
(g) The acquisition time of a sample/hold is ....
(h) In a sampling and digitisation system it can sometimes be advantageous to use a tracking converter in preference to a successive approximation converter because ....
(i) An integrating A/D converter can be used in a sample data system only when ....
(j) In order to avoid aliasing errors in a sampled data system the sampling frequency must be ....
(k) A semiconductor memory designated as a FIFO memory is one in which ....
(l) In a data acquisition system which uses a successive approximation converter without a sample/hold the allowable analog signal bandwidth may be limited by .... rather than by the sampling frequency.

2. Name three functional element performance errors which can influence the accuracy of a data acquisition system.

3. Give three basic factors which may be used in a first design specification for a data acquisition system.

Pick out the statements which are true in questions 4, 5, and 6.

4. A multi-channel data acquisition system which uses analog signal multiplexing rather than digital signal multiplexing
   (a) requires fewer system elements,
   (b) is capable of a faster channel throughput rate,
   (c) is more suitable for use with analog signals derived from several remote transducers.

5. Increase in a sample/hold capacitor value
   (a) decreases the time for which the sample/hold can accurately hold,
   (b) increases the acquisition time of the sample/hold,
   (c) increases the aperture time of the sample/hold.

6. The aperture time of a sample/hold
   (a) depends mainly on the value of the sample/hold capacitor,
   (b) depends on the rate at which the switch resistance changes when the sample/hold is instructed to hold,

(c) is the time taken for the sample/hold to switch from the hold to the sample mode,

(d) is the time delay between the application of the hold command and the time at which the output is no longer affected by changes in the input signal.

7. A sampled data system used for the digital transmission of analog signals uses 8-bit converters. Assuming ideal components what signal to noise ratio do you expect to obtain
   (a) with a full range sinusoidal input signal,
   (b) with a sinusoidal signal whose peak-to-peak value is half the full scale input range?

8. In the system shown in figure 5.5 the sample/hold has an acquisition time of 4.2 μs and the maximum clock frequency for which the system will operate is 2 mHz. What is the highest sampling frequency which can be obtained with the system when using all 8 bits?

9. A sequentially accessed semiconductor memory is made with eight parallel 256-bit shaft registers and is used in a system which produces a delay in an analog signal (see figure 5.13). Find the delay if the analog signal is sampled at a frequency of 100 kHz.

10. A 512-bit shift register and an 8-bit serial in parallel out shift register are used as a serial delay line. The delay line is used to delay an analog signal which is digitised by an 8-bit A/D converter. What is the delay if the sampling frequency is 50 kHz?

11. A system based on that shown in figure 5.15 is to be used to acquire a digital record of the variations of an analog signal which occur in a time period of 5 ms. What sampling frequency must the system operate at if the parallel shift registers which constitute the digital memory each have 512 bits? What bandwidth limit should be placed upon the analog signal? (Neglect slew rate limitations.)

12. A 10-bit successive approximation A/D converter has a full scale range (full scale negative to full scale positive) of 20 V and a conversion time of 12 μs. The converter is used to digitise an analog signal. Find the maximum rate at which the analog signal can change without impairing conversion accuracy
    (a) when the converter is used without a sample/hold,
    (b) when the converter is preceded by a sample/hold module whose acquisition time T_{aq} = 4 μs and aperture time = T_{ap} = 100 ns. What bandwidth limit is it advisable to set for a full scale sinusoidal signal in the two cases?

13. The sampling systems given in question 12 are used in an analog multiplexed

eight-channel data acquisition system. The time taken for the multiplexer to switch channels and for its output to settle is $T_{mux} = 3$ μs. Find the maximum channel throughput rate and the highest allowable frequency for a full range sinusoidal input signal when the converter is used
(a) without the sample/hold,
(b) with the sample/hold.

6 Practical Considerations

Previous chapters have directed attention towards the functional operations performed by data converters, and some of their applications in signal measurement and processing have been discussed. In this chapter we gather together some of the factors requiring the attention of the engineer who is faced with the task of implementing a practical converter application. It is basically a question of selecting a device or devices capable of fulfilling the performance requirements of the application and then using them in such a way that they do in fact attain that desired performance.

Device selection can be a time-consuming task and choice of converter device is usually just one of the many design decisions which must be made by the system designer. A converter application can, in most cases, be satisfactorily implemented with a variety of different devices. It is the designer's task to select devices which provide the most cost-effective solution to the application. Often it is not simply the most cost-effective solution to the converter application but rather the most cost-effective overall system design which is sought.

Converters may be just one of many functional elements in a system and if so the compatibility of the converter device with other system elements must be an important element of any criteria for converter selection. Compatibility relates to device power supply requirements, to logic levels and to ease of interface of the converter with other system elements. For example, a DAC may require an input data latch (see section 2.7) in order to interface with other system elements; a single packet DAC device which incorporates such a latch obviously facilitates interfacing. An A/D converter may be required to interface with a microprocessor system; a converter which incorporates a tri-state digital output latch and whose control signals (start conversion, conversion complete) are compatible with the microprocessor makes for a convenient interfacing. Many of the newer converter devices that are appearing on the market are designed specifically for ease of interface with a microprocessor. Data converter users need to keep up to date with device availability.

6.1 DESIGN PROCEDURES

There is unfortunately no single well-defined procedure which can be followed to guarantee the successful design and practical implementation of an electronic system. However, experience indicates that the most cost-effective systems are achieved as a result of a detailed and systematic approach to the design process. Careful planning and an attention to detail at the initial stages of a design can minimise design errors and inadequacies which otherwise may only show up when a system is nearing completion. Inadequacies discovered in a partially complete system are usually expensive to rectify. They can even involve having to scrap the system and start again!

The amount of design planning required to implement a converter application is clearly dependent on the complexity of the application. However, whether the design be for a complete data acquisition system or a single converter circuit, mistakes are most likely to be avoided if a conscious effort is made to formulate and put down on paper the various stages of the design process.

The realisation of a successful design depends ultimately on the ability to spot all relevant variables and make provision for them. Design planning may be summarised as a two-stage process

Stage 1 Complete statement of application objectives.

Stage 2 Device evaluation and selection with the purpose of fulfilling the application objectives.

Device selection decisions are seldom clear cut. As a general rule the greater the speed and accuracy requirements of a converter application the more costly will be the devices which are needed to fulfil them. Costs are likely to increase exponentially as state-of-the-art performance limits are approached. Economic considerations may dictate design compromises in the form of a relaxation of the speed and/or accuracy of the design specifications as set out in the application objectives.

6.2 APPLICATION OBJECTIVES

This should be as complete a statement as possible of the exact function to be performed by the application, how accurately and under what conditions that function is to be performed. An organised approach to specifying application objectives can be realised by drawing up an application check list. The items which must be included in such a list naturally depend on the specific nature of the application. Certain general considerations which relate to D/A and A/D converter applications can be recognised.

6.2.1 General Considerations for D/A Converter Applications

(1) Resolution—how many bits are there in the input digital data words?

(2) What conversion accuracy is required?
(3) How are the digital data words coded? (See chapter 1).
(4) What are the logic levels of the digital data?
(5) Is the digital data in serial or parallel form? Possible requirements of serial to parallel conversion? (See section 2.7.1.)
(6) What is the nature of the load which is to be supplied by the analog output signal? Load impedance? Voltage or current drive? The desired full scale analog output?
(7) Is the DAC to be operated with a fixed or variable reference? If a fixed reference, is the reference to be internal or external? If a variable reference multiplying DAC application in how many quadrants must the multiplication be valid? (See section 4.3.1.)
(8) Dynamic considerations—how quickly must the system operate? How quickly must the DAC output settle to its desired accuracy after an input data change? Are switching transients (glitches) likely to be of importance? (See section 2.8.)
(9) Under what environmental conditions must the DAC operate to its specified speed and accuracy? What are the allowable errors over the temperature range, supply voltage range, time between recalibration? Are there any particularly unfavourable environmental conditions: lack of space, vibration and shock, high humidity, etc?
(10) Are there any special interface requirements with other system elements? Must the converter inerface with a microprocessor? (See section 5.10.)

6.2.2 General Considerations for A to D Converter Applications

(1) What is the range of the analog input signal?
(2) What is the nature of the analog input signal? Is it slowly varying, rapidly changing, sampled or noisy? Is analog signal processing to be used prior to conversion? Possible use of a differential measurement amplifier to reduce interference noise?
(3) What input impedance must the converter have?
(4) How accurately must the system operate?
(5) How many bits are required?
(6) Type of digital coding required?
(7) How quickly must the converter operate? Conversion time? Conversion time requirements will be a major factor governing the type of conversion technique (see chapter 3).
(8) Under what environmental conditions must the converter operate to its specified speed and accuracy? To what extent are the various error sources tolerable over the expected temperature range, supply voltage range, time between recalibration, etc.? Possible unfavourable environmental conditions?
(9) Is the converter to be operated with a fixed reference? Is the reference to be external or an integral part of the converter package? What are the

Practical Considerations 179

requirements on the reference stability? Is a ratiometric conversion required (See section 3.12.)

(10) Details of control signal requirements: conversion complete, start conversion.

(11) Interface requirements—is an output data latch required? Must the converter interface readily with a microprocessor? If so, which microprocessor? (See section 5.10.)

(12) Is the converter to form part of a multiplexed data acquisition system? If it is, the requirements of the converter will be influenced by the configuration of the other functional elements in the system. Performance requirements of multiplexers, sample/holds and analog conditioning circuitry will need to be considered in assessing allowable error contributions. (See sections 5.8, 5.9.)

Above are some of the points which will require consideration when specifying a converter application. They cannot be regarded as complete lists. The designer, with hindsight, usually finds that his initial design specification is incomplete. The point to be stressed is that the more thought that goes into the initial design specification the less time is likely to be required in modifications at a later and more expensive stage.

The designer should make every effort to anticipate and provide for possible future problems. He should attempt to envisage the occurrence of possible fault conditions which might lead to system malfunction or possible device destruction. Preventive measures should be built into the design. Future demands on the system should, if possible, be anticipated. For example, in specifying a data acquisition system the number of channels is a basic design specification. The designer knows how many channels he needs now but is he sure that there might not be additional parameters which will require measurement? How easily can the number of channels in the system be expanded? Is it better to specify those extra channels in the initial design?

6.3 DEVICE SELECTION

A design objective as specified by an application check list gives a set of criteria for device selection. Device performance parameters provide a means of assessing the ability of specific devices to fulfil design criteria.

Device performance parameters indicate the way in which device behaviour differs from its ideal functional mode. Differences from ideal behaviour represent performance errors which need to be evaluated in order to match devices to required design objectives.

Some performance errors are stated more or less directly by device data sheets. There are other errors which can arise as a result of interactions between system components—they can usually be inferred from data sheet information. Such errors occur mainly in analog signal paths: for example, input impedance of a device causing source loading error, input bias currents of a device giving rise to offset errors when flowing through series resistance in the signal path.

Errors which cannot be predicted from data sheet information can arise because of parasitic interactions (stray capacity, ground loops) between system components. It is the designer's task to minimise such stray coupling mechanisms by careful attention to system layout and the adoption of appropriate grounding and shielding techniques.

6.3.1 Some of the Choices Available [28, 31, 32]

Data conversion products were once the province of a limited number of specialist firms. Specialist firms continue to produce high performance converter devices but the advent of monolithic integrated circuit converter devices has seen the entry of the large multi-national semiconductor firms into the conversion device field.

The user of data conversion products has a choice between devices based on different circuit techniques. The choice is between discrete component modules, hybrid integrated circuit packages and monolithic integrated circuit devices. Specialist converter firms normally offer devices which are based on all three techniques. The big semiconductor manufacturers tend to concentrate their efforts on developing low cost monolithic devices aimed at the large number of converter applications where requirements are in the lower end of the performance range.

Traditionally data converters have been of the discrete component type: first available in instrument cases, later in compact encapsulated modules. The discrete component approach allows the optimum combination of components of all types and has over the years allowed the development of very precise high speed converter modules. Discrete component modules are generally more expensive than hybrid and monolithic devices.

Hybrid converters are made by interconnecting a number of monolithic chips so as to provide a compact single packet functional device. The hybrid approach is almost as flexible as the discrete component approach in that it allows an optimum combination of monolithic chips. The limits on the hybrid approach are set by the availability of suitable monolithic chips and the necessity to minimise the number of chips used in the hybrid package. Decreasing chip count reduces the number of interconnection bonds between chips which in turn reduces labour costs and leads to greater yield and device reliability.

Monolithic integrated circuit converter devices are the least expensive. The first complete D/A converter to be made in monolithic form was produced in 1970 by Precision Monolithics (DAC-01, a 6-bit unit including an internal reference and output amplifier). There have been continual developments and improvements in monolithic converter devices since then. There are a large number of 8-bit D/A and A/D monolithic converters available and an increasing number of higher resolution (more bits) devices are now appearing on the market. The older monolithic device tend to require additional external parts,

Practical Considerations   181

for example, resistors, capacitors, references, operational amplifier comparators, clock circuits. However, many of the newer monolithic devices which are now being produced require very few external components—all necessary system elements are formed on the one monolithic chip. Monolithic converter devices are beginning to take over tasks which formerly required the use of more expensive hybrid or discrete component designs.

One of the main areas of application for A/D converters is in data acquisition systems. In such systems, converters are combined with other functional modules such as multiplexers, sample/holds, instrumentation amplifiers, filters, etc. see section 5.8). The engineer has the choice of assembling a data acquisition system from individual functional modules or he can buy a ready built system. Ready built systems which condition, multiplex, digitise and interface analog input signals to a computer or microcomputer are called *analog input systems*. Also available are ready built systems which interface a computer's digital data to analog output signals through D/A converters—they are called *analog output systems*. Rapid developments are taking place in microcomputer input/output systems. They first became available in the form of single self-containing circuit boards but single packet input/output systems in both hybrid and monolithic integrated circuit form are now available. The cost of ready built input/output systems tends to be considerably more than the parts cost of equivalent systems assembled from individual functional elements. However, if a ready built system is available which meets the requirements of the application objectives it may turn out to be the most cost-effective approach to the system design. Ready built systems provide considerable savings in design and engineering time and in general are less demanding of engineering 'know-how' than user-assembled systems.

The above represents a very brief general review of converter device availability. The reader may well complain about the lack of specific information about particular devices. A reference to specific devices has been deliberately avoided. In a rapidly changing field such as this the only really satisfactory sources of information about what actual devices are available and what they can do are manufacturers' up to date product guides and device data sheets. Successful device selection then stems from an understanding of the meaning of data sheet performance specifications and an ability to interpret those specifications in terms meaningful to a particular application.

6.4 UNDERSTANDING CONVERTER PERFORMANCE SPECIFICATIONS

It can be a tricky business sorting through performance specifications in order to select a converter device for use in an application. There is as yet no accepted standard of converter specifications agreed between all manufacturers. It cannot even be assumed that all the converters produced by the same manufacturer have their performance specified in exactly the same way.

A converter, although it may be in the form of a functional unit, is in itself a complex system whose performance depends on the characteristics of its circuit elements. Since available converter units differ considerably in their state of completeness, that is, in their requirement for external parts, references, operational amplifiers, etc., the absence of standardised specifications is perhaps not very surprising. Fortunately, most converter specifications are fairly self-descriptive—the reader who has followed the text thus far should not have too much difficulty in interpreting them. Often a stumbling block for the newcomer to converters is simply an unfamiliarity with converter terminology. The discussion which follows is intended to overcome this difficulty by bringing together the main terms used to describe converter performance. It should enable the reader readily to extract the information that he needs from converter data sheets.

Converter terminology may be roughly divided into two kinds. Firstly there are terms descriptive of an ideal conversion technique or conversion function which are not specific to a particular converter device. Secondly there are terms which are used to specify the performance of particular converter devices.

In the non-specific terms those descriptive of conversion techniques are tracking converter, successive approximation converter, flash encoder and dual-slope converter (see chapter 3). Terms descriptive of an ideal conversion function are resolution, quantisation uncertainty, scale factor and terms used to describe different types of conversion codes (see chapter 1).

In the second kind of terms are those which describe the speed of operation of particular devices. Also there are terms like offset error, scale error, linearity, monotonicity and missing codes. These terms specify the extent to which particular devices fail to perform an ideal conversion function; they are used to evaluate the errors to be expected when using the devices. Error evaluations allow the matching up of devices to design objectives.

6.5 IDEAL CONVERSION FUNCTIONS

An ideal linear D/A conversion function is described by an equation of the form (see chapter 2)

$$\text{analog output} = \text{scale factor} \times \text{numerical fraction determined by digital input code} \quad (6.1)$$

In an ideal linear A/D conversion function the analog range is divided by evenly spaced levels (quantised). The same digital output code is assigned to all analog input values in the quantisation range between adjacent levels. Analog input values which lie at the centre of each quantisation rage are related to the digital output code assigned to each range by an equation of the form

$$\text{analog input value at centre of quantisation range} = \text{scale factor} \times \text{numerical fraction determined by digital output code} \quad (6.2)$$

Practical Considerations

Digital codes are in the form of a sequence of digital bit values. The values of the numerical fractions in equations 6.1 and 6.2 are related to digital bit values in a manner which depends on the particular conversion code in use.

Assuming $x_1, x_2, x_3, \ldots, x_n$ represent digital bit values, x_1 = most significant bit, MSB, x_n = least significant bit, LSB and x_i = 0 or 1, the values of the numerical fraction in commonly used conversion codes are as follows.

6.5.1 Unipolar Codes

Natural binary

$$\text{Numerical fraction} = x_1 2^{-1} + x_2 2^{-2} + \ldots + x_n 2^{-n} \quad (6.3)$$

Binary coded decimal BCD

most significant quad ↓

$$\text{Numerical fraction} = 0.1\,(x_1 2^3 + x_2 2^2 + x_3 2^1 + x_4 2^0)$$
$$+ 0.01\,(x_5 2^3 + x_6 2^2 + x_7 2^1 + x_8 2^0)$$
$$+ 0.001\,(x_9 2^3 + x_{10} 2^2 + x_{11} 2^1 + x_{12} 2^0) + \text{etc.} \quad + \text{etc.}$$
$$(6.4)$$

Note that in a BCD quad only the first ten of a natural binary sequence of code words are valid.

6.5.2 Bipolar Codes

Offset binary

$$\text{Numerical fraction} = \left(x_1 2^{-1} + x_2 2^{-2} + \ldots + x_n 2^{-n}\right) - \tfrac{1}{2} \quad (6.5)$$

Two's complement

$$\text{Numerical fraction} = \left(\bar{x}_1 2^{-1} + x_2 2^{-2} + x_3 2^{-3} + \ldots + x_n 2^{-n}\right) - \tfrac{1}{2}$$
$$(6.6)$$

Symmetrical offset binary

$$\text{Numerical fraction} = 2\left(x_1 2^{-1} + x_2 2^{-2} + \ldots + x_n 2^{-n}\right) - \frac{2^n - 1}{2^n} \quad (6.7)$$

Sign magnitude

Numerical fraction = $\left(x_2 2^{-1} + x_3 2^{-2} + \ldots + x_n 2^{-(n-1)} \right)$

when $x_1 = 1$ and

$$- \left(x_2 2^{-1} + x_3 2^{-2} + \ldots + x_n 2^{-(n-1)} \right) \quad (6.8)$$

when $x_1 = 0$.

Ideal D/A and A/D conversion functions using a 3-bit natural binary code are illustrated graphically in figure 6.1.

6.5.3 Resolution and Quantisation Uncertainty

Discontinuous incremental changes are inherent in any counting system. Numbers change in steps—the smallest change in a numerical magnitude is governed by the weighting associated with the lowest order digit in the number. On the other hand, analog variables, by their very nature, are capable of a continuous variation. There is no lower limit to the smallest change possible in an analog variable. It follows that there is a limit to the precision (a resolution limit) with which an analog magnitude can be prescribed numerically. Whenever a numerical measure of magnitude is assigned to an analog variable there is an element of uncertainty even though the measurement system be ideal. The uncertainty is $\pm \frac{1}{2}$ the weighting associated with the lowest order digit in the number.

A D/A converter produces an analog output signal prescribed by an input digital code word (by a binary number). There is a limit to the resolution with which the digital code can prescribe the analog signal magnitude. An A/D converter produces a binary number which represents the magnitude of an analog input signal. There is inevitably, even in an ideal conversion, an element of uncertainty in the binary number. Analog values which lie in the uncertainty range all give rise to the same binary number and are not resolved as different.

An ideal D/A converter using an n-bit binary code can produce 2^n discrete analog output values corresponding to the 2^n different states of its input code (see equations 6.1, 6.3, 6.5, 6.6, 6.7 and 6.8). Output values are separated by an amount equal to the LSB weighting which is nominal full scale range/2^n. Full scale range is the difference between maximum and minimum analog values, a unipolar range 0 to 10 V or a bipolar range -5 V to $+5$ V both constitute a 10 V full scale range. The relative value of the LSB, that is its value expressed as a fraction of nominal full scale range is the *resolution* of the DAC. Resolution of an n-bit binary DAC is variously expressed as one part in 2^n, as a percentage $(1/2^n) \times 100$ per cent or sometimes simply as n bit. Resolution of an A/D converter is expressed in the same way.

It should be noted that data converters always have their analog range expressed as nominal full scale rather than actual full scale. Nominal full scale remains the same regardless of the number of bits. The output of a DAC can never span its nominal full scale range and an A/D converter does not have

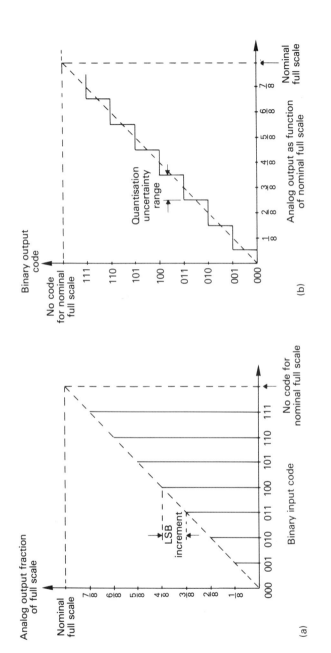

Figure 6.1 Ideal 3-bit binary DAC and ADC transfer function: (a) DAC transfer function; (b) ADC transfer function

output codes assigned to all of its nominal full scale range. The reason for this apparent discrepancy is not difficult to see. An n-bit code has 2^n states but there are only $(2^n - 1)$ steps separating the states. Actual full scale range is therefore always one LSB increment less than nominal full scale range.

Quantisation uncertainty is a term which is strictly speaking applicable to A/D converters but not to DACs. It is the uncertainty associated with an A/D converter's finite resolution. All analog values within a quantum range are designated by the same digital code value. There is therefore an inherent element of uncertainty in this code value. In an ideal A/D conversion relationship the quantisation range is symmetrical about exact code values (see figure 6.16) and there is a quantisation uncertainty of $\pm \frac{1}{2}$ LSB. An n-bit binary converter has a quantisation uncertainty $\pm \frac{1}{2}$ full scale range$/2^n$.

Resolution and quantisation uncertainty both depend on the number of states allowed in the conversion code. Converters using a BCD code have, for the same number of bits, a smaller number of allowed code states than converters which use a binary code. A 12-bit binary code has 2^{12} states; it gives a resolution $(1/2^{12}) \times 100$ per cent = 0.024 per cent and quantisation uncertainty range $\pm \frac{1}{2}$ full scale range$/2^{12}$. A 12-bit BCD code has only 1000 states; it gives a resolution 0.1 per cent and quantisation uncertainty range ± 0.05 per cent full scale range.

It should be remembered that finite resolution and quantisation uncertainty are characteristics of ideal conversion functions. Performance deficiencies can cause the separation of some of the adjacent analog output levels produced by a practical DAC to be greater than the 'ideal' separation full scale range$/2^n$, and the uncertainty in an A/D converter's output digital code can be greater than the ideal $\pm \frac{1}{2}$ LSB quantisation uncertainty.

A distinction is sometimes made between the *nominal resolution* and the *useful resolution* of a practical DAC device. Nominal resolution is as defined above. Useful resolution is the actual (as distinct from ideal) biggest separation of adjacent analog output levels over the full range of code values and operating conditions expressed as a fraction of nominal full scale. Alternatively useful resolution is expressed as the smallest uniquely distinguishable bit number over all operating conditions. For example, a 12-bit converter has a nominal resolution of one part in 2^{12}, $(1/12^{12}) \times 100$ per cent but because of non-linearities its useful resolution over its operating temperature range may only be one part in 2^{10}, $(1/2^{10}) \times 100$ per cent, a 10-bit useful resolution. Data sheets do not normally contain an explicit statement of useful resolution but its value over a particular range of operating conditions can normally be inferred from linearity specifications.

6.6 ERROR SPECIFICATIONS

One of the main factors governing the selection of a data converter is that its performance meets the accuracy requirements as set down in the application

Practical Considerations

objectives. Real converters do not perform the ideal conversion functions which were discussed in the p̩ vious section. The extent to which a converter's performance departs from an ideal conversion function is described by various performance parameters which are to be found in a device data sheet. Departures from ideal behaviour represent performance errors. The designer must interpret data sheet performance parameters in such a way as to evaluate the errors which are likely to occur when using a specific device in a particular application. In this way he may match a device to the accuracy requirements of the application.

The reader should be aware of the convention often adopted of quoting accuracy specifications in terms of error specification. For example, 0.1 per cent accuracy really means 0.1 per cent inaccuracy or error and the accuracy should strictly be quoted as 99.9 per cent accurate. Data converter specification sheets seldom include an overall accuracy specification. Overall accuracy is governed by an accumulation of separate errors. Individual error sources are normally specified separately and further treatment of accuracy is deferred until after these separate error contributions have been identified and discussed.

There are three basic error sources in a data converter: offset error, 'gain' or scale factor error and linearity error. These error sources are all present simultaneously and all exhibit a dependence on temperature and device supply voltage; their values also change (drift) over the long term with time. Variation of error sources with temperature are normally the most significant variation. Two of the error sources, offset error and gain error, can usually have their initial values trimmed to zero by the converter user but their variations remain. Linearity error (strictly non-linearity error) cannot be trimmed. If offset and gain errors were zeroed and a device maintained at constant temperature and used with constant supply voltages device non-linearity would remain to limit conversion accuracy. The only way to achieve small non-linearity errors is to choose a device which inherently exhibits small departures from linearity.

6.6.1 Offset (Zero) Error

An ideal DAC produces zero analog output when the digital bit values assigned to the code representing zero are applied to its digital inputs. Any analog output other than zero under these output conditions is called the offset or zero error of the DAC.

In an A/D conversion analog values within one of the partitions of the analog continuum are all designated by the single digital code word which is assigned to represent zero. The mean of the analog input range required to produce the zero digital output code is zero in an ideal A/D conversion. Any mean analog input other than zero required to produce this code is called the offset or zero error of the A/D converter.

DAC and ADC transfer functions exhibiting offset error are shown graphically in figure 6.2. Offset error in a DAC is often due to the input offset error of the DAC's output operational amplifier. In a feedback A/D converter system (see

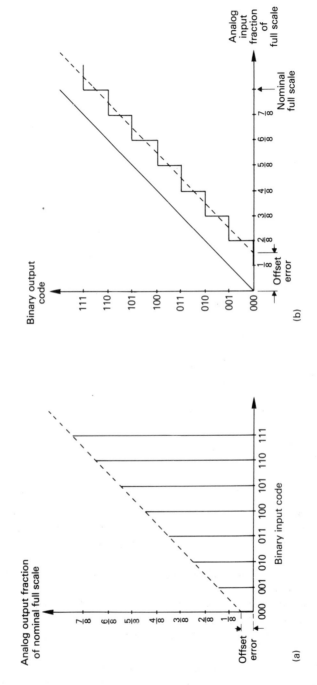

Figure 6.2 3-bit binary DAC and ADC transfer function with offset errors: (a) DAC transfer function with offset error; (b) ADC transfer function with offset error

Practical Considerations 189

chapter 3) the offset error of the DAC used in the system causes the A/D converter to have an offset error. The system's comparator can introduce an additional offset component. Offset errors in integrating-type A/D converters are due to offsets in the systems analog conditioning circuitry. Integrating A/D converters often incorporate an *auto-zero* function whereby the converter's input offset error is adjusted to zero by the action of a feedback loop. The *auto-zero* function is included in a conversion cycle : offset errors are zeroed prior to each conversion.

6.6.2 Gain Error (Scale Factor Error)

The 'gain' or scale factor of a data converter is the number which establishes the relationship between the converters analog values and digital code fractions (see equations 6.1 and 6.2). Gain or scale error is the amount by which the converter's actual scale factor differs from its designed or nominal value. It is usually expressed as a percentage deviation from the nominal value.

Gain determines the slope of a converter's transfer function as shown in figure 6.3. In the transfer functions shown in figure 6.3 the conversion functions have a slope which differs from the desired value. The gain error may be defined as the difference in full scale values between the ideal and actual transfer functions when the offset error is zero (expressed as a percentage of the designed nominal full scale).

A converter's gain or scale factor normally depends on reference voltage and scaling resistor values (see chapters 2 and 3). Scale factor can be set to a desired value by adjustment of these parameter values. The adjustment must be performed after trimming out offset error; the converter's actual full scale analog value is then adjusted to be one LSB less than the desired nominal full scale value.

6.6.3 Linearity

The ideal conversion functions embodied in equations 6.1 and 6.2 are linear. In a graphical representation of the input/output relationship implemented by an ideal DAC a straight line may be drawn passing through all the DAC's analog output values (see figure 6.1a). An ideal A/D converter's input/output graph is similarly characterised by a straight line passing through the analog input values which lie at the mid points of each quantisation range (see figure 6.1b). The transfer functions implemented by real converters exhibit departures from these ideal straight line plots. In this section we discuss the terms commonly used to describe and specify a converter's non-linearity. A converter's linearity specifications are of critical importance since errors due to non-linearity cannot be adjusted, unlike offset and gain errors which can be trimmed to zero by the converter user.

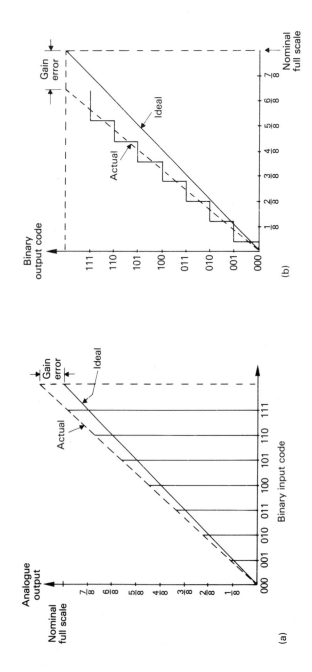

Figure 6.3 3-bit binary DAC and ADC transfer functions with gain errors: (a) DAC with gain error; (b) ADC with gain error

Practical Considerations 191

Conventionally a curve's non-linearity is often expressed in terms of its deviation from a 'best fit' straight line, this being the straight line from which the curve deviates least. A best fit straight line specification of a converter's non-linearity is, however, not convenient for the converter user. The location of a converter's transfer function on an input/output set of axes is controlled by two adjustments. An offset error trim procedure shifts the transfer function parallel to the axis to make it pass through zero. Adjustment of a converter's scale factor rotates the transfer function to allow trimming out the converter's gain error—the adjustment is normally performed at full scale. These two adjustments fix the end points of the converter's transfer function. In order to fit in with this calibration procedure, converter manufacturers normally provide an end point straight line specification of a converter's non-linearity.

A converter's *linearity error* or non-linearity is specified as the maximum deviation of the converter's transfer function from a straight line passing through zero and full scale under the conditions in which the converter's offset and full scale gain errors have been adjusted to zero. Linearity error is generally expressed in LSBs or as a percentage of the converter's full scale range. An end point specification of non-linearity is a conservative one: it can mean an actual linearity which is twice as good as that specified in terms of a best bit straight line (see figure 6.4). Determining a best fit straight line is a tedious process when calibrating a converter. It involves a knowledge of the 'shape' of the converter's non-linearity in order that scale trimming can be performed at the analog value

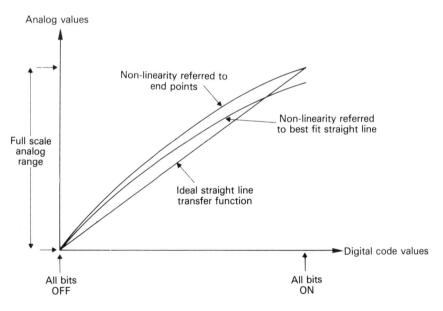

Figure 6.4

which locates the transfer function so that its deviation from the best fit straight line is divided equally above and below the line. It is for this reason that most converter manufacturers have opted for an end point specification of non-linearity.

Linearity error as defined above is sometimes referred to as *integral linearity error*. Integral linearity error gives the maximum deviation of a converter's transfer function from an ideal straight line but it says nothing about the small scale nature of any non-linearities. Non-linearities in the form of deviations from the ideal LSB analog increment change between adjacent steps in a converter transfer function are described by a specification which is called differential linearity error.

Differential linearity error is the deviation from the ideal LSB (full scale range/2^n) analog difference between any two adjacent code values in the complete range of the converter. If a converter has a specified differential linearity error of $\pm \frac{1}{2}$ LSB then the actual size of any quantum step in its transfer function is never less than $\frac{1}{2}$ LSB and never more than $1\frac{1}{2}$ LSB.

Differential linearity error in a DAC transfer function is due to errors in the weighting of the DAC's bit current increments. Consider a simple example as an illustration. A 3-bit natural binary coded DAC has the MSB current increment $\frac{1}{4}$ LSB high and bit 2 current increment $\frac{1}{4}$ LSB low. The DAC produces analog output values in proportions $0:1:1.75:2.75:4.25:5.25:6:7$ instead of the ideal proportions $0:1:2:3:4:5:6:7$. The DAC's transfer function would be as shown in figure 6.5a. The biggest differential linearity error occurs at the code transition 011 to 100, where the DAC output changes by $1\frac{1}{2}$ LSB instead of the ideal 1 LSB change; this represents a differential linearity error of $\frac{1}{2}$ LSB. If this DAC were to be used as part of a feedback A/D converter system (say a successive approximation A/D converter, see chapter 3) there would be a corresponding differential non-linearity in the A/D converter's transfer function as shown in figure 6.5b.

Arising from differential linearity errors there are two other terms used in describing converter transfer functions. The term monotonicity applies to DACs and the term missing or skipped code applies to A/D converters. A *monotonic* DAC is one whose analog output continuously increases (or remains the same but never decreases) as its digital input code is incremented up. The transfer function shown in figure 6.5a is monotonic.

A DAC may be *non-monotonic* if its differential linearity error is greater than ± 1 LSB. Figure 6.6a shows a non-monotonic transfer function. In the code transition 011 to 100 the DAC's output goes down instead of up. The differential linearity error is greater than 1 LSB at this region of the transfer function. In the DAC transfer function shown in figure 6.6b there is a differential linearity error greater than 1 LSB yet monotonicity is maintained. A differential linearity error specified as greater than ± 1 LSB does not necessarily imply non-monotonicity but it can allow it. In order to guarantee monotonicity the differential linearity error must be less than ± 1 LSB.

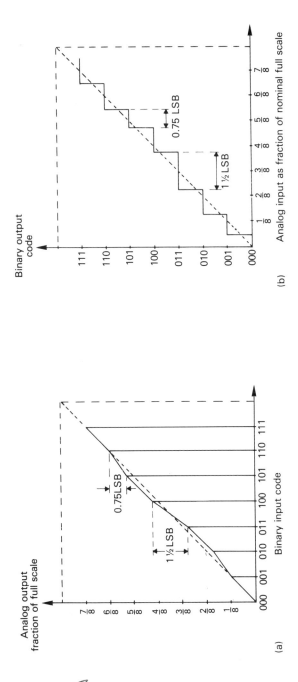

Figure 6.5 Differential linearity error: (a) DAC with differential linearity error; (b) ADC with differential linearity error

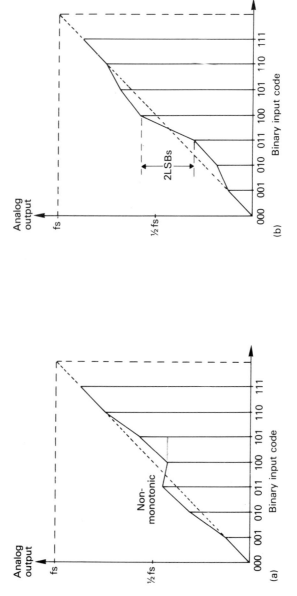

Figure 6.6 DAC transfer functions with differential linearity error greater than ± 1 LSB: (a) non-monotonic transfer function; (b) monotonic transfer function

Practical Considerations

The biggest differential linearity errors in a DAC's transfer function occur at major code transitions. In a multi-bit DAC these are the transitions such as those at $\frac{1}{4}, \frac{1}{2}, \frac{3}{4}$ full scale. In the transition which occurs at $\frac{1}{2}$ full scale the MSB current increment is switched on and all lower order bits are switched off. Consider, for example, the case of a 12-bit DAC. Its MSB current increment is 2^{11} times bigger than its LSB current increment. An error any greater than $(1/2^{11}) \times 100 \triangleq 0.05$ per cent in the MSB current increment can cause a differential linearity error greater than 1 LSB at the $\frac{1}{2}$ scale transition.

The predominant kind of non-linearity in A/D converters of the feedback type (for example successive approximation A/D converters, see chapter 3) is differential linearity error. It arises because of the differential linearity error of the DAC used in the A/D converter system. Differential linearity error in an A/D converter's transfer function manifests itself in an uneven spacing of the quantisation uncertainty ranges which exist at each code value (see figure 6.5b). If a non-monotonic DAC is used in a successive approximation A/D converter system, the system will have *missing or skipped codes*. One or more of the theoretically possible output code states of the A/D converter will never appear no matter what the value of the analog input signal. In figure 6.7 we show the type of

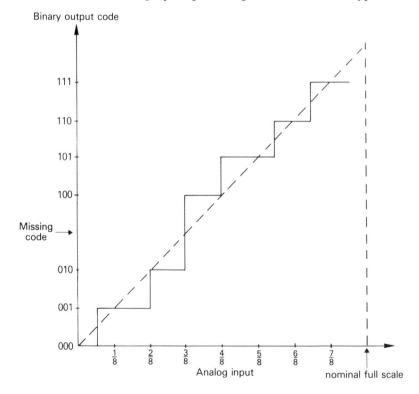

Figure 6.7 Successive approximation A/D converter using non-monotonic DAC has missing code

transfer function to be expected in a successive approximation A/D converter using the non-monotonic DAC whose transfer function was illustrated in figure 6.6a. Note that in figure 6.7 there is no value of the analog input signal for which the output code 011 is produced. 011 is a missing code in this transfer function. Differential linearity errors must be less than ± 1 LSB in order to guarantee no missing codes in a A/D converter's transfer function. Differential linearity errors greater than ± 1 LSB can cause missing codes.

Differential linearity error is almost non-existent in integrating-type A/D converters (for example dual-slope converters, see chapter 3). Any curvature in an integrating A/D converter's transfer function is caused by non-linearities in the analog circuitry (for example, a non-ideal integrator) and is adequately described by the converter's integral linearity specification. A possible source of differential linearity error in an integrating A/D converter would be a variation in the time period of the clock pulses. However, changes in clock frequency which may take place during a conversion are negligible and differential linearity error is therefore negligible in integrating A/D converters.

In summary, a converter's integral linearity error specifies the maximum deviation of the converter's transfer function from an ideal straight line. The specification applies under the conditions in which the converter's offset error and full scale gain error have been zeroed. Integral linearity error includes differential linearity error, that is, it includes both the effects of non-linearities in analog circuitry and the effects of errors in bit current weighting. Differential linearity error describes the deviation from an ideal step size in a converter's transfer function. Differential linearity error must be less than ± 1 LSB to guarantee monotonic operation of a DAC or no missing codes in an A/D converter. An integral linearity error of less than $\pm \frac{1}{2}$ LSB ensures a differential linearity error of less than ± 1 LSB and guarantees monotonicity and no missing codes.

6.6.4 Errors due to Temperature Change

Change in ambient temperature cause changes in a converter's offset gain and linearity error. Offset change with temperature is usually specified in μV per °C or in parts per million of full scale range per °C. Gain change with temperature is specified as full scale temperature coefficient or gain drift in parts per million per °C and linearity error change in parts per million of full scale range per °C. Alternatively, *integral* linearity error and differential linearity error may be specified as a worst case error over a defined temperature range. It is advisable to perform offset and gain error trims with the converter at the temperature at which it will normally operate. (The temperature inside an instrument case can be considerably higher than the temperature outside.) Temperature-induced offset and gain error will then not include the effect of the initial case warm up.

Linearity error cannot be trimmed out by the converter use. A converter's differential linearity error temperature coefficient needs particular attention. A converter may be monotonic (no missing codes in the case of an A/D converter)

Practical Considerations

at normal temperature but because of differential linearity error temperature coefficient may go non-monotonic if the temperature goes outside a specified range. For example, assuming an initial maximum differential linearity error of $\pm\frac{1}{2}$ LSB, for monotonicity to be maintained the change in differential linearity error over temperature range must be no more than an additional $\frac{1}{2}$ LSB. In the case of a 12-bit converter $\frac{1}{2}$ LSB is $\frac{1}{2}\times$ full scale range/2^{12}, that is, 120 ppm of full scale. If the operating range is say $0°C$ to $70°C$ and the converter's initial linearity specs are given at $25°C$, maximum temperature change is $-25°C$ or $+45°C$. The converter's differential linearity error temperature coefficient must be less than $120/45 = 2.7$ ppm of full scale per $°C$ to guarantee monotonicity in the case of a DAC or no missing codes in the case of an A/D converter.

6.7 DYNAMIC RESPONSE PARAMETERS

A finite time is required for a converter to establish a correspondence between its analog and digital signals. Time-dependent parameters have been treated in earlier chapters. A DAC's time response is governed by its *settling time* (see section 2.8). *Conversion speed* or *conversion time* specify the time taken by an A/D converter to perform a conversion—it is the time which elapses between the application of a start conversion command and the time at which the converter signals conversion complete. Throughput rate is another term which is applied to A/D converters and data acquisition systems. *Throughput rate* is the number of digital sample values that can be taken per second. It depends on the conversion time and the time required to set up ready for the next conversion. In a data acquisition system the set up time is a composite of various time delays for example multiplexer, settling time, sample/hold acquisition time (see section 5.9.2).

Conversion speed is the main factor governing the selection of an A/D converter by choice of conversion technique (see section 3.13).

6.8 ACCURACY

A converter's ability to meet a specified accuracy objective in an application is one of the main factors governing its selection. In order to assess this ability the converter's performance parameters, which are published by the manufacturer, must be used to predict the maximum error likely to occur when using the converter in the application. Sheets seldom include an overall accuracy specification because of the multiplicity of factors governing a converter's accuracy data. The designer has to compute overall accuracy by means of a systematic error budget analysis. An error budget uses data sheet information to identify and assess the magnitude of all possible error sources in the known range of operating conditions. If the individual errors are unrelated to one another the worst possible error and hence the overall accuracy figure is then obtained by simply adding the separate error contributions.

Care is required in interpreting accuracy specifications and in relating device performance parameters to errors. When comparing the accuracy of converters made by different manufacturers it must not be assumed that they mean exactly the same thing when they publish identical numbers for a particular parameter. Manufacturers' literature must be studied carefully in order that any differences in their definitions of terms may be identified and interpreted.

Newcomers to converters often confuse the terms resolution and accuracy. A finite resolution (see section 6.5.3) is a property of an ideal error-free converter but a practical converter has an accuracy limit which is determined by its performance errors. Converter manufacturers differ somewhat in their use of the term accuracy and our present usage of the term therefore requires clarification.

The accuracy (strictly inaccuracy error) of a DAC will be taken to mean the maximum deviation of a DAC's output from its designed value, over the full range of operating conditions, expressed as a percentage of the DAC's designed nominal full scale output range. This definition of accuracy includes the effects of initial offset, gain and linearity errors and also their drift components. A DAC's resolution together with its scale factor (see section 6.5) fixes what the DAC's output levels ought ideally to be. A DAC's accuracy error gives the maximum deviation of the actual output levels it produces from these ideal levels.

A clear distinction between resolution and accuracy is perhaps not so easily made for A/D converters as it is for DACs. Even an ideal A/D converter, in a sense, is never quite 'accurate': it has a quantisation uncertainty or error because of its finite resolution. Some manufacturers, when referring to the accuracy of a practical A/D converter, include the effect of quantisation error together with offset, gain and linearity errors. Other manufacturers exclude quantisation uncertainty from the error contributions to accuracy and prefer to regard accuracy as expressing how closely an A/D converter's performance approaches the ideal.

In this discussion the term accuracy as applied to an A/D converter will be interpreted as follows. The inaccuracy error of an A/D converter will be taken to mean the maximum difference between the nominal value of analog input assigned to any output code and an actual value of analog input which will produce that code, the error to be expressed as a percentage of the converter's nominal full scale input range. The nominal value of analog input assigned to an output code is that which is determined by the converter's ideal conversion equation which is of the form

analog input = scale factor × fraction determined by output code

Our use of the term accuracy applied to an A/D converter includes quantisation error together with all other error sources due to non-ideal performance. It is clearly not possible in our usage of the term to set a measure of the inaccuracy associated with an A/D converter which has missing codes.

6.8.1 Error Analysis

In a converter error analysis it is usual to express all error components in terms of a percentage of the converter's full scale range. Error terms which are normally included are as follows.

(1) Initial errors (normally specified at 25°C)
These include initial offset gain and linearity errors. A calibration procedure is often performed prior to using a converter in order to set the initial offset and gain errors to zero. The procedure, in effect, fixes the end points of the converter's transfer function at their desired positions on the input/output axes. Some converter manufacturers call a converter accuracy at the instant of calibration its relative accuracy. On this usage of the term, relative accuracy is determined only by the converter's linearity error, for the one performance error which remains at the instant of calibration is the converter's linearity error. (It should be noted that all converter manufacturers do not attribute exactly this meaning to the term relative accuracy.)

(2) Accuracy drift with temperature
Error terms are obtained by multiplying offset, gain and linearity temperature coefficients by the expected change in device temperature. Care is required in estimating the accuracy drift of bipolar converters because the offset drift and gain drift errors due to the reference voltage temperature coefficient may not be independent errors (see example 1b below).

(3) Error due to power supply changes
Error is obtained by multiplying the specified full scale range sensitivity (percentage change of full scale range per percentage change of supply voltage) by the expected percentage change in power supply voltages.

(4) Error due to long term drift with time
Not all data sheets provide information about drift with time. Time drift errors are not likely to be significant in the lower resolution converter applications (say 8 bits). In high resolution applications the designer should obtain information about a converter device's long term stability.

The sum of the above error terms gives the worst case error. The actual error in a converter application is likely to be less than this worst case error since it is unlikely that all errors will be in the same direction. RMS addition of errors gives a smaller resultant error, but this is probably too optimistic because the number of error sources is not large.

6.8.2 Examples of Error Evaluations

Examples of worst case error evaluations are now given. In an error analysis the method used to estimate error contributions has to be adapted to the nature of the device error data available.

The following error data is available for an 8-bit current output DAC (Precision Monolithies DAC-08H). Specifications apply for $V_s = \pm 15$ V; $I_{ref} = 2.0$ mA $T_A = 0°C$ to $70°C$.

Non-linearity	$T_A = 0$ to $70°C$	$\pm 0.1\%$ fs
Full scale temperature coefficient	TC I_{fs}	± 50 ppm/°C (max)
Zero scale current	I_{ZS}	$1.0 \mu A$ (max)
Full range current (All bits ON)	I_{FR} $T_A = 25°C$	$(\frac{255}{256} \times 2 \pm 0.4\%)$ mA
Power supply Sensitivity	PSS I_{fs+} PSS I_{fs-}	$\pm 0.01\%/\%$ (max) $\pm 0.01\%/\%$

Performance data for a voltage reference, an operational amplifier and a comparator are also available.

Voltage reference output voltage $V_o = 10$ V $\pm 0.5\%$

Output voltage temperature coefficient TC $V_o = 20$ ppm/°C

Operational amplifier input offset voltage $V_{io} = 5$ mV

Input offset voltage temperature coefficient $\dfrac{\Delta V_{io}}{\Delta T} = 20 \mu V/°C$

Input bias current $I_B = 40$ nA

Bias current temperature coefficient $\dfrac{\Delta I_B}{\Delta T} = 1$ nA/°C

Comparator input offset voltage $V_{io} = 1$ mV

Offset voltage temperature coefficient $\dfrac{\Delta V_{io}}{\Delta T} = 1 \mu V/°C$

Bias current $I_B = 500$ nA

Bias current temperature coefficient $\dfrac{\Delta I_B}{\Delta T} = 5$ nA/°C

Worst case error evaluations are to be made for the following systems implemented with the above devices.

(1) Voltage output DAC
(a) unipolar full scale range 10 V;

(b) bipolar (offset binary code) full scale range −5 V to +5 V.
(2) Successive approximation A/D converter
(a) unipolar full scale input range 10 V;
(b) bipolar (offset binary code) full scale input range −5 V to +5 V.

In each case an estimate of the worst case error prior to calibration is required. Initial offset and gain errors are then to be assumed zeroed at 25°C and the worst case accuracy over the temperature range 0°C to 70°C is to be evaluated. Power supplies are assumed to have a maximum 1 per cent variation. Resistor values with 0.5 per cent tolerance are to be used.

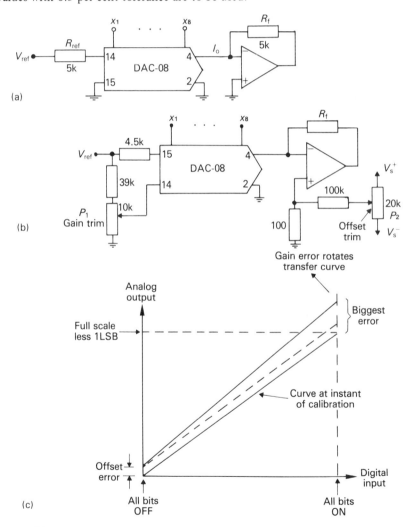

Figure 6.8 Unipolar voltage output DAC (example 1a): (a) basic voltage output DAC; (b) calibration arrangements; (c) errors exclusive of linearity error

Example 1(a)
The basic voltage output DAC system with no adjustment circuitry is shown in figure 6.8a (see section 2.4). The initial error at 25°C is made up of the initial offset error, the initial gain error and the DAC's linearity error.

Initial offset This is due to operational amplifier initial offset error and I_{ZS}, the DAC's output leakage current, which is present when all bits are OFF. The normal method of evaluating an operational amplifier's offset error is used. Total operational amplifier input offset voltage is

$$E_{os} = V_{io} + I_B R_f + I_{ZS} R_F$$
$$= 5 \times 10^{-3} + 40 \times 10^{-9} \times 5 \times 10^3 + 10^{-6} \times 5 \times 10^3$$
$$= 10.2 \text{ mV}$$

The operational amplifier is connected as a current to voltage converter and since the DAC-08 has a high output impedance the feedback fraction β may be taken as unity. The input offset error appears at the operational amplifier output multiplied by $1/\beta$ ($1/\beta = 1$). Thus

$$\text{worst case initial offset error} = \pm \frac{10.2 \times 10^{-3}}{10} \times 100 = \pm 0.1\% \text{ of full scale}$$

Initial gain error The 'gain' is nominally

$$S = \frac{V_{ref}}{R_{ref}} R_f$$

Initial gain error consists of V_{ref} tolerance ± 0.5 per cent, R_f tolerance ± 0.5 per cent, R_{ref} tolerance ± 0.5 per cent and the DAC full range current tolerance ± 0.4 per cent. The worst case initial gain error is ± 1.9 per cent.

Linearity error This is specified as ± 0.1 per cent. The worst case error (the 'accuracy') prior to calibration is thus $0.1 + 1.9 + 0.1 = \pm 2.1$ per cent. Although the initial offset error of ± 0.1 per cent is perhaps tolerable the initial gain error of ± 1.9 per cent is clearly not tolerable.

Suitable offset and gain trimming arrangements are shown in figure 6.8b. Offset is trimmed first: with all bits OFF potentiometer P_2 is adjusted to make the output of the operational amplifier zero. Scale trim is performed by switching all bits ON and adjusting P_1 to make the operational amplifier output voltage 9.961 V (full scale less one LSB).

Errors after calibration Offset error becomes the offset drift over the maximum temperature change 45°C (70 − 25°C).

$$\text{Offset drift error} = \frac{\Delta V_{io}}{\Delta T} \times 45 + \frac{\Delta I_B}{\Delta T} R_f \times 45 + I_{ZS} R_f$$

Practical Considerations

$$= 20 \,\mu V/°C \times 45°C + 1 \,nA/°C \times 5 \,k\Omega \times 45 + 1 \,\mu A \times 5 \,k\Omega$$

$$= 6.125 \,mV$$

$$\approx \pm 0.06 \text{ per cent of full scale}$$

Note that no data is available for $\Delta I_{ZS}/\Delta T$; we assume a worst case drift over temperature range of 1 μA for I_{ZS} (clearly pessimistic).

Gain error over temperature is due to TC I_{fs} of $\pm$ 50 ppm/°C and V_{ref} temperature coefficient 20 ppm/°C. An additional gain drift due to differential temperature tracking of R_{ref} and R_f is possible.

$$\text{Gain error} = \pm \frac{50}{10^6} \times 100 \times 45\% \pm \frac{20}{10^6} \times 100 \times 45\%$$

$$= 0.23\% + 0.09\%$$

$$= 0.32\%$$

Error due to power supply changes ± 0.01 per cent due to V_s^+ change and ± 0.01 per cent due to V_s^- change; total $= \pm 0.02\%$.

Linearity error over temperature is 0.1 per cent.

Total worst case error ('accuracy') after calibration

$$= \pm 0.06\% \pm 0.32\% \pm 0.02\% \pm 0.1\%$$

$$= \pm 0.5\% \text{ of full scale}$$

In this example the biggest single error contribution is that due to the gain drift with temperature. In order perhaps to avoid unnecessary overspecification it should be realised that while offset error applies to all outputs (it translates the transfer curve) gain error is essentially a percentage of output (it rotates the transfer curve). The effect of offset and gain curves on the transfer curve are shown exclusive of linearity error in figure 6.8c. The largest possible errors occur at full scale.

Example 1(b)

Offset binary operation is obtained by injecting a half full scale output current into the summing point of the DAC's output operational amplifier; the basic arrangement is illustrated in figure 6.9a. With resistor R_1 in circuit the effect on errors of V_{ref} tolerance and temperature coefficient is not the same as in the unipolar circuit. Also R_1 makes the output amplifier's feedback fraction $\beta = R_1/(R_1 + R_f)$ and the amplifier's input offset error appears at its output multiplied by $1/\beta = 1 + R_f/R_1$ (= 1.5 in this example).

The operational amplifier acts as an inverter on V_{ref} and with all bits OFF the amplifier produces an output voltage

$$V_o = -V_{ref} \frac{R_f}{R_1}$$

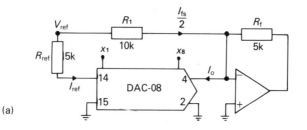

(a)

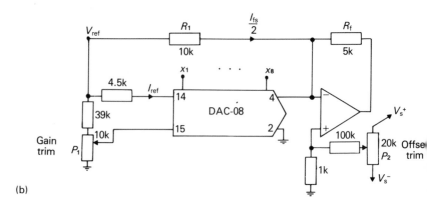

(b)

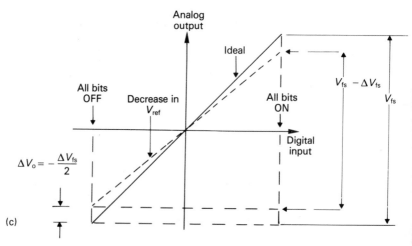

(c)

Figure 6.9 Offset binary bipolar voltage output DAC error evaluations (example 1b): (a) basic offset binary voltage output DAC; (b) offset binary voltage output DAC with offset and gain trims; (c) error in bipolar transfer curve due to reference tolerance or temperature coefficient

Practical Considerations

R_1 is proportioned so as to give the required half full scale range offset—in this example $R_1 = 2R_f$.

A departure ΔV_{ref} from its nominal value, either due to initial tolerance or temperature coefficient causes an error

$$\Delta V_o = -\Delta V_{ref} \frac{R_f}{R_1} \tag{6.9}$$

in the converter's full scale negative output voltage. The error is of opposite polarity to the error in V_{ref}.

V_{ref} sets the value of the converter's full scale output range through the relationship

$$V_{fs} = \frac{V_{ref}}{R_{ref}} R_f$$

An error ΔV_{ref} in V_{ref} causes an error in full scale range

$$\Delta V_{fs} = \frac{\Delta V_{ref}}{R_{ref}} R_f \tag{6.10}$$

Equations 6.9 and 6.10 give

$$\Delta V_{fs} = -\Delta V_o \frac{R_1}{R_{ref}} = -2\Delta V_o \tag{6.11}$$

V_{ref}-induced offset and full scale range errors are of opposite polarity. Their combined effect on the converter's transfer function is shown in figure 6.9c. Errors which are equal in magnitude but of opposite polarity are produced at full scale negative and full scale positive. There is no offset of the bipolar converter's output at zero output. The full scale errors expressed as a percentage of full scale range (-5 V to $+5$ V) are

$$\frac{\Delta V_{ref} R_f / R_1}{V_{ref} R_f / R_{ref}} \times 100\% = \frac{\Delta V_{ref}}{V_{ref}} \frac{R_{ref}}{R_1} \times 100\%$$

$R_1 = 2R_{ref}$ making the errors

$$\pm \tfrac{1}{2} \frac{\Delta V_{ref}}{V_{ref}} \times 100\% \tag{6.12}$$

Initial errors in the bipolar configuration are therefore as follows.

Initial offset error (exclusive of reference) The operational amplifier's initial input offset error E_{os} multiplied by $1/\beta$

$$E_{os} = V_{io} + I_B R_f \| R_1 + I_{ZS} R_f \| R_1$$
$$= 5 \times 10^{-3} + 40 \times 10^{-9} \times 3.3 \times 10^3 + 10^{-6} \times 3.3 \times 10^3$$
$$= 8.43 \text{ mV}$$

Offset error is thus 8.43 × 1.5 = 12.6 mV. Expressed as a percentage of full scale range this is ± (12.6/10 × 10⁻³) × 100 ≙ 0.13 per cent fs. Offset error due to R_f and R_1 tolerance is ± 0.5 per cent full scale.

Initial gain error (exclusive of V_{ref} tolerance) R_{ref} tolerance ± 0.5 per cent, R_f tolerance ± 0.5 per cent and DAC full scale range tolerance ± 0.4 per cent. Worst case value ± 1.4 per cent full scale range.

Error due to V_{ref} tolerance

$$\pm \tfrac{1}{2} \frac{\Delta V_{ref}}{V_{ref}} \times 100 = \pm 0.25\%$$

Linearity error This is, specified as ± 0.1 per cent. The worst case error (the 'accuracy') of the bipolar converter prior to calibration is thus

± (0.13 + 0.5 + 1.4 + 0.25 + 0.1) = ± 2.4% fsr   (fsr = 10 V)

Both gain and offset trims are clearly desirable in the bipolar configuration in order to bring the initial error to a tolerable level. Offset and gain trimming arrangements are shown in figure 6.9b. An alternative offset adjustment would be to trim resistor R_1. It should be realised that a potentiometer connected to the operational amplifier summing point increases the capacitance at the amplifier summing point which may adversely effect the amplifier's transient behaviour.[1]

Offset trim should be performed at full scale negative, with all bits OFF the offset potentiometer P_2 adjusted to make the operational amplifier output −5 V. Gain trim is performed by switching all bits ON and adjusting potentiometer P_1 to make the operational amplifier output +4.961 V (full scale positive − 1 LSB).

Errors after calibration   Offset drift over temperature

$$45 \times \frac{\Delta E_{os}}{\Delta T} = \frac{\Delta V_{io}}{\Delta T} \times 45 + \frac{\Delta I_B}{\Delta T} \times 45 \times R_f \| R_1 + I_{ZS} R_f \| R_1$$

$$= 20\, \mu V/°C \times 45° + 1\, nA/°C \times 45 \times 3.3\, k\Omega + 1\, \mu A \times 3.3\, k\Omega$$

$$= 0.9 + 0.148 + 3.3 = 4.35\, mV$$

Offset temperature drift error $= 4.35 \times \dfrac{1}{\beta} = 4.35 \times 1.5 = 6.53\, mV$

$$= \frac{6.53 \times 10^{-3}}{10} \times 100 = 0.065\% \text{ fs range}$$

Gain error over temperature (exclusive of V_{ref} temperature coefficient)

$$50\, ppm/°C \times 45 = \frac{50}{10^6} \times 45 \times 100$$

$$\triangleq \pm 0.23\%$$

Effect of V_{ref} temperature coefficient (equation 6.12)

$$\frac{\Delta V_{ref}}{V_{ref}} \times \tfrac{1}{2} \times 100\% = \frac{20 \times 45}{10^6} \times \tfrac{1}{2} \times 100$$

$$= \pm 0.045\%$$

Power supply sensitivity

± 0.02%

Linearity error over temperature

± 0.1%

Total worst case error (accuracy) after calibration is

± (0.065 + 0.23 + 0.045 + 0.02 + 0.1) ≙ ± 0.46% fs range

Note full scale range is −5 V to +5 V = 10 V.
The biggest single error contribution is gain drift over temperature. In the bipolar configuration gain error produces the biggest errors at full scale positive the smallest error at full scale negative.

Example 2(a)
A basic unipolar successive approximation A/D converter system is shown in figure 6.10a. Errors prior to calibration are as follows.

Offset error is due to comparator offset and I_{ZS}

± $(V_{io} + I_B R_{in} + I_{ZS} R_{in})$ = ± (1 mV + 500 nA × 5 kΩ + 1 μA × 5 kΩ)

= ± 8.5 mV = ± 0.085% fs

Gain error is due to tolerance in V_{ref}, R_{in}, R_{ref} and DAC full scale current

± (0.5 + 0.5 + 0.5 + 0.4) = ± 1.9%

Linearity error

± 0.1%

Quantisation uncertainty

−1 LSB (no $\tfrac{1}{2}$ LSB offset) = −0.39%

Total worst case error prior to calibration = ± (0.085 + 1.9 + 0.1) −0.39

= 1.695% to −2.475%

Offset and gain trim circuitry is included in figure 6.10b. The offset adjustment is performed first. The input signal V_{in} is set at 19.5 mV (0 +$\tfrac{1}{2}$ LSB) and potentiometer P_1 is adjusted to make the output code dither between 00000000 and 00000001. The input signal is then set at 9.941 V (full scale − 1.5 LSB) and

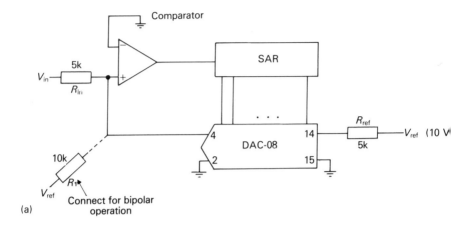

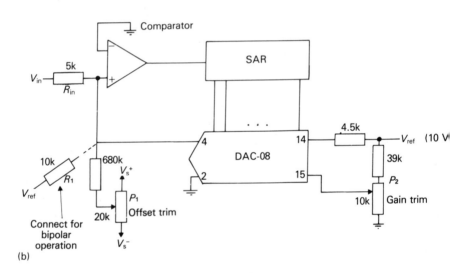

Figure 6.10 Successive approximation A/D converter (example 2): (a) basic successive approximation A/D converter; (b) successive approximation A/D converter with offset and gain trims

the gain trim potentiometer P_2 is adjusted so as to make the output code dither between the values 11111110 and 11111111.

Errors after calibration   Offset drift over temperature

$$\pm \left(\frac{\Delta V_{io}}{\Delta T} \times 45 + \frac{\Delta I_B}{\Delta T} \times R_{in} \times 45 + I_{ZS} \times R_{in} \right)$$

assumed maximum change

$= \pm (1\,\mu\text{V}/°\text{C} \times 45°\text{C} + 5\,\text{nA}/°\text{C} \times 5\,\text{k}\Omega \times 45°\text{C} + 1\,\mu\text{A} \times 5\,\text{k}\Omega)$

$= \pm 5.17\,\text{mV} = \pm 0.0517\%\,\text{fs}$

Gain drift over temperature is due to V_{ref} temperature coefficient $\pm 20\,\text{ppm}/°\text{C}$ and TC $I_{\text{fs}} \pm 50\,\text{ppm}/°\text{C}$

$$\pm \left(\frac{20}{10^6} \times 45 \times 100 + \frac{50}{10^6} \times 45 \times 100 \right) = \pm 0.32\%\,\text{fs}$$

Power supply sensitivity (DAC)

$\pm 0.02\%$

Linearity error (DAC)

$\pm 0.1\%$ over temperature

Quantisation uncertainty

$\pm \frac{1}{2}\,\text{LSB} = \pm 0.19\%\,\text{fs}$

The total worst case error ('accuracy') of the unipolar A/D converter after calibration is

$\pm (0.052 + 0.32 + 0.02 + 0.01 + 0.19) = \pm 0.68\%\,\text{fs}$

Example 2(b)
Bipolar operation with a full scale input range $-5\,\text{V}$ to $+5\,\text{V}$ is obtained by injecting a current equal to half the DAC's full scale output current into the comparator summing point. This is accomplished by resistor R_1 connected to V_{ref} as indicated in figure 2.10. In the bipolar circuits offset and full scale range errors due to error in V_{ref} are similar to those for the bipolar DAC (see figure 6.9c). Errors of equal magnitude but opposite sign are produced at full scale negative and full scale positive but there is no offset at zero scale. The size of the error is

$$\pm \tfrac{1}{2} \frac{\Delta V_{\text{ref}}}{V_{\text{ref}}} \frac{R_{\text{ref}}}{R_1} \times 100$$

(see equation 6.12).

Offset error (exclusive of V_{ref} tolerance) is due to comparator offset

$\pm (V_{\text{io}} + I_B R_{\text{in}} \parallel R_1 + I_{\text{ZS}} R_{\text{in}} \parallel R_1)$

$= \pm (1\,\text{mV} + 500\,\text{nA} \times 3.3\,\text{k}\Omega + 1\,\mu\text{A} \times 3.3\,\text{k}\Omega)$

$= \pm 5.85\,\text{mV} = \pm 0.058\%\,\text{fs}$

Offset due to R_{in} and R_1 tolerance

$\pm 0.5\%\,\text{fs}$

Gain error (exclusive of V_{ref} tolerance) is due to R_{ref} tolerance ± 0.5 per cent, R_{in} tolerance ± 0.5 per cent, DAC full range tolerance ± 0.4 per cent

total ± 1.4%

Error due to V_{ref} tolerance

$$\pm \tfrac{1}{2} \frac{\Delta V_{ref}}{V_{ref}} \times 100\% = \pm 0.25\%$$

Linearity error (DAC)

± 0.1%

Quantisation uncertainty

− 1 LSB (no $\tfrac{1}{2}$ LSB offset)

− 0.39% fs

The total worst case error (accuracy) of the bipolar converter prior to calibration is

± (0.058 + 0.5 + 1.4 + 0.25 + 0.1) − 0.39

= + 1.92% to −2.7% fs (−5 V to +5 V = 10 V)

In the bipolar converter offset trim is performed first at full scale negative. The input signal is set at −4.98 V (−5 + $\tfrac{1}{2}$ LSB) and the offset potentiometer P_1 is adjusted so as to make the output code dither between the values 00000000 and 00000001. Gain trim is performed at full scale positive. The input signal is set at +4.94 V (+5 − 1.5 LSB) and potentiometer P_2 is adjusted so as to make the output code dither between the values 11111110 and 11111111.

Errors after calibration   Offset drift over temperature (exclusive of V_{ref} drift)

$$\pm \left(\frac{\Delta V_{io}}{\Delta T} \times 45 + \frac{\Delta I_B}{\Delta T} \times R_{in} \| R_1 \times 45 + I_{ZS} R_{in} \| R_1 \right)$$

= ± (1 µV/°C × 45 + 5 nA/°C × 3.3 kΩ × 45°C + 1 µA × 3.3 kΩ)

= ± 4.1 mV = ± 0.041% fs

Gain drift over temperature (exclusive of V_{ref} drift)

± 50 ppm/°C × 45°C = ± 0.23%

V_{ref} drift over temperature

$$\pm \tfrac{1}{2} \frac{\Delta V_{ref}}{\Delta_{ref}} \times 100 = \pm \tfrac{1}{2} 20 \text{ ppm/°C} \times 45 \times 100\%$$

= ± 0.045%

Practical Considerations

Power supply sensitivity (DAC)

± 0.02%

Linearity error over temperature (DAC)

± 0.1% fs

Quantisation uncertainty

± $\frac{1}{2}$ LSB = ± 0.19% fs

The total worst case error ('accuracy') of the bipolar A/D converter after calibration is

± (0.041 + 0.23 + 0.045 + 0.02 + 0.1 + 0.19) = ± 0.62% fs

Full sale range is −5 V to +5 V = 10 V.

The above examples are illustrative of the type of error budget analysis that a designer is advised to perform when making a choice of devices for use in a data converter application. The examples given show the effects of external devices and components on the converter system errors. Error analyses for self-contained converter systems are usually less involved because manufacturers normally provide overall performance data which is inclusive of the effects of internal references, operational amplifiers, comparators, resistors, etc.

Error budgets show up the most critical performance-limiting parameters in an application; the designer must pay particular attention to these parameters when comparing different devices. Also revealed are those parameters in which a relaxation in specification is perhaps possible to allow the use of less expensive devices.

It should of course be realised that an error budget is only as reliable as the assumptions on which it is based. Particular care must be taken so as not to omit any significant errors. If a converter forms part of a data acquisition or distribution system the converter's error budget is simply a part of the overall system error budget. This must include the possible errors in all system elements: amplifier errors, multiplexer errors, sample/hold errors, etc.

Time-dependent errors are not normally included in an error budget analysis but they must not be forgotten. A DAC's analog output does not immediately respond to a change in its digital input signal but is in error for a finite time; transient errors can be large (see section 2.8: settling time, glitches). The conversion time of a successive approximation A/D converter is a function of the number of bits and the clock frequency (see section 3.6). The use of a higher than recommended clock frequency, in an attempt to decrease conversion time, can introduce additional errors not included in an error budget based on a device's specified performance data. There is usually a speed/accuracy design compromise involved in most data conversion systems (see section 5.9.2).

Yet more errors, not included in an error budget based on specified performance data, are those which can arise because of the way in which devices are connected together and used. Some of these extraneous error sources and ways of combatting them are discussed in the next section.

6.9 APPLYING CONVERTERS

Having taken considerable care to select devices which, according to their published performance specifications, are capable of meeting the speed and accuracy requirements of an application, it is important not to degrade systems performance by improper attention to circuit details external to the devices. The exact way in which devices are connected to each other and to external passive components and power supplies can have a significant effect on system performance. A device's performance parameters are specified under the assumption that a set of definite electrical conditions exists at its external pin connections. If interactions which take place outside a device cause undesired signals to be injected at the pin connections it would not be surprising if the device should apparently fail to achieve its rated performance.

An increasing number of 12-bit converters are appearing on the market and 16-bit devices are now available from several manufacturers. Although engineers may talk quite glibly about such devices they should not forget that the LSB of a 12-bit converter represents only 0.025 per cent of full scale (0.0015 per cent of a 16-bit converter's full scale). Even extremely small extraneous error sources are significant when looking for LSB order accuracies with these devices.

Extraneous error signals are not a problem peculiar to converters—they are troublesome in all sensitive analog systems and are called interference signals or interference noise. If the mechanisms whereby interference error signals are injected into a system are understood, such error signals, although not completely removable, can be considerably reduced.

Circuit techniques for the reduction of interference signals in a system are mainly concerned with the physical layout of the system: the relative positions of the systems functional elements and components and the nature of the interconnections between them. A practical electronic circuit or system has parasitic components which do not appear in its theoretical circuit diagram. Parasitic components are determined by the practical circuit layout: they provide the principal mechanism for the injection of extraneous errors.

Connecting wires or printed circuit boards' etch strips have a finite resistance and inductance which can allow error voltages to be developed along them. Components or conductors which are close to one another have stray capacitance between them. Stray capacitance provides a mechanism for the injection of error signals—particularly vulnerable are high impedance points in analog circuitry. Stray capacitance to ground at a circuit point can slow down the dynamic signal response. Stray capacitance to ground at the summing point of an operational amplifier can have a marked effect on the amplifiers closed loop stability.[1]

Practical Considerations 213

Stray alternating magnetic fields, if they cut across large area circuit loops in an analog circuit induce interference signals into the circuit.

Published performance parameters for available converter devices are attained as a result of their designers' close attention to the minimisation of internal parasitic error interactions. The user of these devices must devote similar care to minimising possible extraneous errors which have their origins outside the devices. Some particular points which should concern the converter user are now described.

(1) Avoid errors due to incorrect grounding.

A 'ground' point in a circuit is a common zero reference point to which signals at other points in the circuit are referred. The separate functional units in a multi-element converter system each have their own separate 'ground' pins. These separate ground pins must be connected together when the system is wired up. The grounding problem is concerned with the manner in which the separate ground pins are tied together.

Problems arise if separate functional units or different sections of a unit share a common return path to ground. In such circumstances ground return currents, flowing in the non-zero impedance of the common path cause an extraneous coupling of signals between the separate sections. The extraneous signals can represent significant errors in high resolution data converter systems (significant, that is, in relation to LSB magnitude). Particularly to be avoided in data conversion systems are errors due to the coupling of digital signals into analog sections of the system.

The user of data converter devices must concern himself more with grounding principles the greater the number of separate units there are in his system which require connecting together. However, even the most self-contained of data converters requires external signal, power supply and ground connections and manufacturers' recommend installation procedures should be followed if they are available.

A/D converter units normally have a number of separate external 'ground' pins (not connected together internally). These ground pins are referred to as digital ground or digital common, analog ground or common and analog input signal ground. Digital ground is the point which all digital currents in the unit return to the digital power supply. Analog common is the point at which analog currents return to the analog power supply. Analog input signal ground is the zero reference point in the system to which analog signals are referred.

The principles involved in wiring up an A/D converter unit are illustrated in figure 6.11. In the wiring arrangement shown in figure 6.11a all ground terminals are connected together as they must be, but in such a way as to introduce extraneous error signals. Figure 6.11b shows the correct method of wiring up. In the incorrectly wired arrangement the power supply currents for both the analog and digital sections of the system's circuitry together with any output or display currents all flow through a line common to the input

214 Data Converters

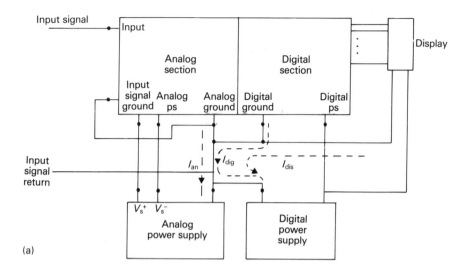

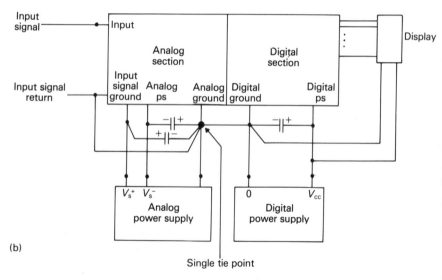

Figure 6.11 Making external connections to an A/D converter unit: (a) incorrectly wired A/D converter unit; (b) correctly wired A/D converter unit

signal. The non-zero impedance of this common lead causes both a d.c. offset error voltage and varying error voltages to be injected at the input to the system. In the correct method of wiring ground pins are tied together at one

Practical Considerations 215

point only and separate ground returns to this common tie point are provided. The common tie point should be made as close to the converter unit as possible, preferably to a large ground plane say under the unit. Power supply leads should be no longer than necessary, supplies should be decoupled with capacitors connected to the units' pin connections as shown, electrolytic capacitors (say 1 μF) should be bypassed with 0.01 μF ceramic types.

In a data conversion system which is assembled using separate IC packages, each IC has its own power supply and ground pins. It is important to distinguish clearly between the different ground pins. In no circumstance should currents returning to digital power supply ground share a common return line with analog signal return paths or with analog power supply return currents. As an example of wiring practice figure 6.12 shows the separate ground connections advisable in the practical sampling and digitisation system which was described in section 5.4 (see figure 5.5).

(2) Avoid capacitance coupling of digital signals into analog lines.

Stray capacity which exists between a conductor carrying a digital signal and a conductor carrying an analog signal provides a mechanism for coupling digital error signals into analog circuitry. In order to minimise this effect digital data and control lines should be routed as far as possible away from lines carrying analog signals.

(3) Locate components sensibly.

In a converter unit assembled with IC packages, a layout which allows the use of short analog leads should be sought. Any leads associated with the comparator summing point in a successive approximation A/D converter should be as short as possible. The location of a ready built converter unit within a system is also important. It should be such as to minimise external interference signals. A converter should not be positioned near to potential sources of electromagnetic interference, for example, not near a transformer or electric motor.

A/D converters should be located as close as possible to their input signal sources. Interference signals between source ground and converter ground can be troublesome when signal source and converter have to be connected by a long cable run. A remedy is to use a differential input measurement amplifier immediately preceding the converter. The long cable run from the signal source then goes to the differential input terminals of the amplifier. The interfering signal becomes a common mode signal and it is rejected to an extent determined by the common mode rejection ratio, CMRR, of the amplifier. [1, 9]

D/A converters should be positioned as close as possible to their loads.

(4) Look out for thermal effects.

Care should be taken not to exceed device power supply and current limits which might lead to excessive internal power dissipation. Internal power dissipation within an IC device can cause a thermal gradient to exist within the chip. Thermal gradients can upset the precise thermal tracking of tran-

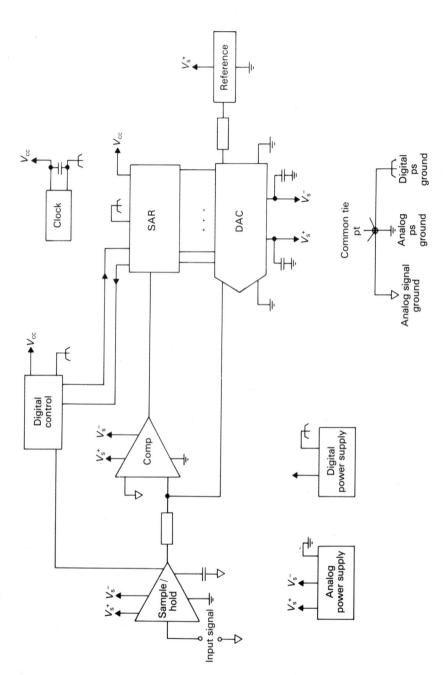

Figure 6.12 Ground and power supply connections in sampling system assembled from ICs.

sistors within a chip that is necessary if a low offset drift with temperature is to be maintained. The thermal tracking of resistors, necessary for low converter gain drift with temperature, can also be upset.

An additional source of error in high resolution converter devices may arise because of thermoelectric voltages. A thermoelectric emf is generated whenever there is a difference in temperature between the junctions of dissimilar metals. Thermal gradients between IC packages or between printed circuit boards can give rise to significant thermoelectric errors.

(5) Use adequate quality external passive components.

External potentiometers which may be used for gain and offset adjustments should be capable of precise setting and should have a low temperature coefficient of resistance. Fixed value resistors which may be wired externally to set gain should be high stability low temperature coefficient types. If internal gain setting resistors are provided in a converter package they should be used to set gain in preference to external resistors. In a DAC or a successive approximation A/D converter which uses a DAC it is the stability of the resistor ratio (operational amplifier feedback resistor or input resistor/resistor used to set reference current) which controls gain stability. Temperature tracking of resistor values is more important than temperature stability of individual values. Internal resistors are likely to thermally track better than resistors mounted outside a converter package.

(6) Perform calibration procedure carefully.

The general principles involved in performing gain and offset trims were discussed in the previous section in connection with the error evaluation examples. If recommended gain and offset adjustment procedures are given by the device manufacturer they should be followed. However, whatever adjustment procedure is adopted it should be remembered that a converter's accuracy after calibration can be no better than the accuracy of the instruments used to fix calibrating voltages.

Adjustment potentiometers should be connected to a converter unit with short leads. They should be mounted in such a way that they are accessible when the converter unit is installed within its instrument case. Calibration should then be performed with the converter at the temperature at which it will normally operate.

(7) Use full scale input signals to an A/D converter.

An A/D converter's accuracy is expressed in terms of a percentage of the converter's full scale input range. If the error is to represent the smallest possible percentage of signal it is necessary to make use of the full input range of the converter. If input signals much less than full scale are expected it is advisable to include an amplifier in the system preceding the converter. Analog signal conditioning with a differential input measurement amplifier can be used as a means of reducing common mode error signals. If input signals of interest span a wide range the possibility of logarithmic compression prior to conversion should be considered (see section 5.6.1).

6.10 SELF-ASSESSMENT EXERCISES

1. Give five items likely to be required in a design checklist for
   (a) a DAC application,
   (b) an A/D converter application.

2. Which of the following terms are descriptive of a converters departure from the ideal?
   (a) non-monotonic
   (b) quantisation uncertainty
   (c) offset binary
   (d) offset error
   (e) integrating
   (f) missing codes.

3. What numerical fraction is represented by the digital word 10010110 in the following codes?
   (a) natural binary
   (b) BCD
   (c) offset binary
   (d) two's complement
   (e) sign magnitude.

4. Which of the following errors can normally be corrected by the converter user?
   (a) offset error
   (b) quantisation error
   (c) scale factor error
   (d) linearity error.

5. Fill in the blanks in the following statements.
   (a) An analog input system is the term normally used to denote . . . .
   (b) Converter manufacturers normally provide an end point straight line specification of a converters non-linearity because. . . . .
   (c) Differential non-linearity in a DAC's transfer function is due to . . . .
   (d) In order to ensure the monotonicity of a DAC transfer function the integral linearity error must be . . . .
   (e) The actual full scale analog range of a converter is always one LSB increment less than its nominal full scale range because . . . .

6. A 12-bit DAC has its differential linearity error specified as $\pm \frac{1}{2}$ LSB at $25°C$. If the differential linearity error temperature coefficient is specified as ± 4 ppm of full scale per $°C$, over what temperature range may the DAC's transfer function be expressed to remain monotonic?

Practical Considerations 219

7. The MSB weighting in a 10-bit natural binary coded DAC is 0.2 per cent less than its correct value. Assuming no error in the other bit weightings, what differential linearity error will be observed in the DAC's transfer function? At which code transaction will the error occur? Will the DAC be monotonic?

8. An A/D converter has the following specification: nominal resolution 12 bits; differential linearity error at 25°C ± ½ LSB; differential linearity error temperature coefficient ± 3 ppm of full scale per °C; gain temperature coefficient ± 25 ppm per °C; offset temperature coefficient ± 5 ppm of full scale per °C; power supply sensitivity 0.002 per cent per per cent. Assuming initial gain and offset errors are trimmed at 25°C, find the worst case conversion error over an operating temperature range of 0 to 50°C. Assume a possible 1 per cent change in power supply.

References

1. G. B. Clayton, *Operational Amplifiers*, 2nd ed. (Butterworth, London, 1979)
2. M. Yuen, D-A converters low-glitch design lowers parts count in graphic displays, *Electronics*, **52**, 16 (1979) 131-5
3. Eugene L. Zuch (ed.), Designing with a sample-hold won't be a problem if you use the right circuit, in *Data Acquisition and Conversion Handbook* (Datel-Intersil, Mansfield, Mass., 1979)
4. Eugene L. Zuch (ed.), Keep track of a sample-hold from mode to mode to locate error sources, in *Data Acquisition and Conversion Handbook* (Datel-Intersil, Mansfield, Mass., 1979)
5. Eugene L. Zuch (ed.), Pick sample-holds by accuracy and speed and keep hold capacitors in mind, in *Data Acquisition and Conversion Handbook* (Datel-Intersil, Mansfield, Mass., 1979)
6. D. Soderquist and J. Schoeff, Low cost, high speed analog-to-digital conversion with the DAC-08, *Precision Monolithics Inc. Full Line Catalog AN-16*
7. Lee, Evans. The intergrating A/D converter, in *Data Acquisition and Conversion Handbook*, ed. Eugene L. Zuch (Datel-Intersil, Mansfield, Mass., 1979)
8. R. Allan, Breaking the data-conversion speed barrier, *Electronics*, **53** (1980) 109
9. G. B. Clayton, *Linear Integrated Circuit Applications* (Macmillan, London and Basingstoke, 1975)
10. J. Schoeff and D. Soderquist. Different and multiplying D to A converter applications, *Precision Monolithics Inc. Full Line Catalog AN-19*
11. Eugene L. Zuch (ed.), Video analog-to-digital conversion, in *Data Acquisition and Conversion Handbook* (Datel-Intersil, Mansfield, Mass., 1979)
12. Eugene L. Zuch (ed.), Put video A/D converters to work, in *Data Acquisition and Conversion Handbook* (Datel-Intersil, Mansfield, Mass., 1979)
13. D. Soderquist, 3 IC 8 bit binary digital to process current converter with 4-20 mA output, *Precision Monolithics Inc. Full Line Catalog AN-21*
14. *Application Guide to CMOS Multiplying D to A Converters* (Analog Devices Inc. Norwood, Mass.
15. J. Edrington, D-a converter forms programmable gain control, *Electronics*, **48**, 15 (1975) 92
16. J. Schoeff and D. Soderqist, DAC-08 applications collection, *Precision Monolithics Inc. Full Line Catalog AN-17*
17. Daniel H. Sheingold (ed.), *Analog-Digital Conversion Handbook* (Analog Devices Inc., Norwood, Mass., 1972)
18. C. Vinn, Digital nulling of OP-05 and SSS725, *Precision Monolithics Inc. Full Line Catalog AB-2*
19. D. Soderquist, Exponentially digitally controlled oscillator, *Precision Monolithics Inc. Full Line Catalog AN-20*
20. J. A. Betts, *Signal Processing, Modulation and Noise* (Hodder & Stoughton, London, 1970)

References

21. Barry A. Blesser, Digitisation of Audio, *J. Audio Engng Soc.*, **26** (1978) 739-71
22. Comdac Companding D/A Converter DAC-76, *Precision Monolithics Inc. Full Line Catalog 1979*
23. Daniel H. Sheingold (ed.), *Nonlinear Circuits Handbook* (Analog Devices Inc., Norwood, Mass., 1974)
24. Y. J. Wong and W. E. Ott, *Function Circuits, Design and Applications*, Burr Brown Electronic Series (McGraw-Hill, New York, 1976)
25. George F. Bryant, Applications of analog conversion products in micro-computers, in *Data Acquisition and Conversion Handbook*, ed. Eugene L. Zuch, (Datel-Intersil, Mansfield, Mass., 1979)
26. T. M. Frederiksen and J. B. Cecil, C-MOS d-a converters match most microprocessors, *Electronics*, **53**, 14 (1980) 140-44
27. D. Fullagar, P. Bradshaw, L. Evans and W. O'Neill, Interfacing data converters and microprocessors, *Electronics*, **49**, 25 (1976) 81-9
28. L. Mattera, Data converters latch onto microprocessors, *Electronics*, **50**, 18 (1977) 81-90
29. M. Tuthill and D. P. Burton, Low cost a-d converter links easily with microprocessors, *Electronics*, **52**, 18 (1979) 149-54
30. D. Lewis, Anatomy and applications of the analogue microprocessor, *Electron. Prod. Des.*, **1** (1980) 40-5
31. N. Mokhoff, Monolithic approach bears fruit in data conversion, *Electronics*, **52**, 10 (1979) 105-16
32. R. Allan, The inside news on data converters, *Electronics*, **53**, 16 (1980) 101-12

Answers to Self-assessment Exercises

CHAPTER 1

1. (a) unit of information; (b) two ... two separate voltage levels; (c) group of bits; (d) one at a time, ...all at the same time; (e) $2^5 = 32$.

2. (b), (e), (f)

3. (a) 9, 23, 6; (b) $\frac{1}{2} + \frac{1}{4} = \frac{3}{4}$, $\frac{1}{8}$, $\frac{1}{4}$, $1/2^6 = 1/64$, $\frac{1}{4} + 1/16 = 5/16$

4. The digital word applied to the DAC is

    MSB   LSB
    1 0 1 0 0 1

(i) (a) Using a natural binary code 101001 gives analog output $(\frac{1}{2} + \frac{1}{8} + 1/64) \times$ full scale $= 0.641 \times 10 = 6.41$ V.
(b) Offset binary converter code: $+ (1/4 + 1/32) \times$ full scale $= 0.28 \times 10 = 2.8$ V. Note that the MSB carries the sign information, the remaining 5 bits indicate the magnitude.
(c) In two's complement the MSB gives sign information. MSB = 1 corresponds to negative values. For negative values the 'magnitude bits' (01001) follow a natural binary complementary code. Thus we have $-$ [full scale $- (1/4 + 1/32)$] full scale $= - 23/32$ full scale $= - (23/32) \times 10 = - 7.2$ V
(ii) (a) Natural binary DAC: magnitude of analog voltage has maximum value $63/64 \times 10 = 9.84$ V for digital input 111111. Offset binary: magnitude of analog voltage has maximum value of full scale $= 10$ V ($-$fs) for digital input 000000. Two's complement: magnitude of analog voltage has maximum value of full scale $= 10$ V ($-$fs) for digital input 100000.
(b) In all cases the smallest analog output corresponds to one LSB of magnitude information. In the natural binary code all 6 bits carry magnitude information in the bipolar codes only 5 bits carry magnitude information. Thus for natural binary we have: minimum analog output $= 1/64$ fs $= 10/64 = 0.16$ V produced by digital input 000001. For offset binary: minimum analog output $= 1/32$ fs $= 0.31$ V produced by digital inputs 100001 (+ve), 011111 ($-$ve). For two's

Answers to Self-assessment Exercises   223

complement: minimum analog output = 1/32 fs = 0.31 V produced by digital inputs 000001 (+ve), 111111 (−ve).

5. There is an uncertainty in the digital output (quantisation uncertainty) of $\pm \frac{1}{2}$ LSB. This is a 3-bit A/D converter; the LSB corresponds to the fraction $1/2^3 = 1/8$. Let x_1, x_2, x_3 represent the digital output states where x_1 is the MSB, x_3 the LSB and $x_i = 0$ or 1. The bit values must be such that they satisfy the relationship

$$\left(\frac{x_1}{2} + \frac{x_2}{4} + \frac{x_3}{8} + \frac{1}{16}\right) > \frac{V_{in}}{\text{full scale}} > \left(\frac{x_1}{2} + \frac{x_2}{4} + \frac{x_3}{8} - \frac{1}{16}\right)$$

that is

$$\left(\frac{x_1}{2} + \frac{x_2}{4} + \frac{x_3}{8} + \frac{1}{16}\right) > 0.7 > \left(\frac{x_1}{2} + \frac{x_2}{4} + \frac{x_3}{8} - \frac{1}{16}\right)$$

The inequality is satisfied by $x_1 = 1, x_2 = 1, x_3 = 0$ that is

$$\frac{1}{2} + \frac{1}{4} + \frac{1}{16} > 0.7 > \frac{1}{2} + \frac{1}{4} - \frac{1}{16}$$

$$0.79 > 0.7 > 0.69$$

Thus voltage levels at X, Y, Z are: X = +5 V, Y = +5 V, Z = 0.

6. (i) (a) 4-bit natural binary conversions
(b) 4-bit offset binary conversions
(c) BCD conversions

(ii) The digital output code 0111 corresponds to the analog input signals. For natural binary conversion

$$\left(\frac{7}{16} \pm \frac{1}{32}\right) 10$$

that is, the input range 4.69 to 4.06 V. For offset binary conversion

$$-\left(\frac{1}{8} \pm \frac{1}{16}\right) 10$$

that is, the input range −1.88 to −0.63 V. For BCD conversion

$$\left(\frac{7}{10} \pm \frac{1}{20}\right) 10$$

224 Data Converters

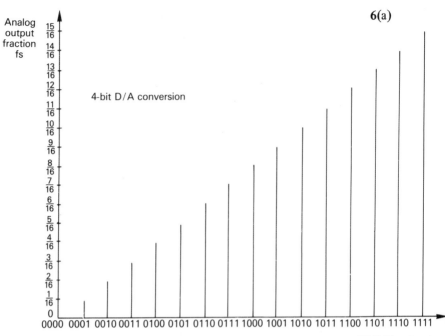

6(a)

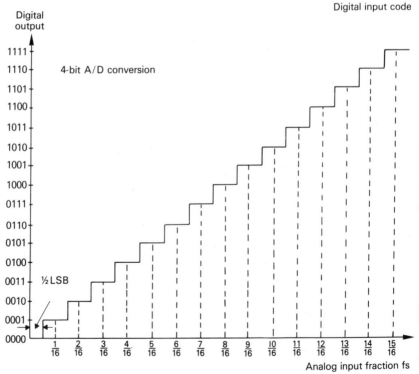

Answers to Self-assessment Exercises

6(b)

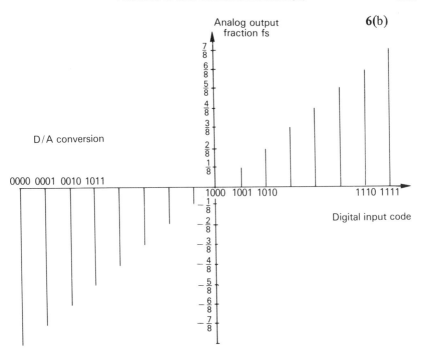

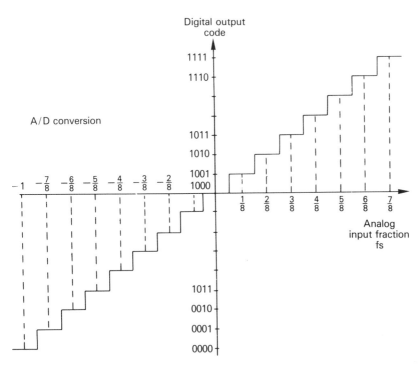

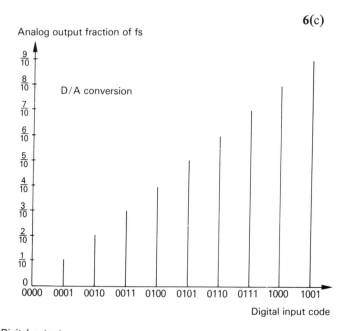

6(c)

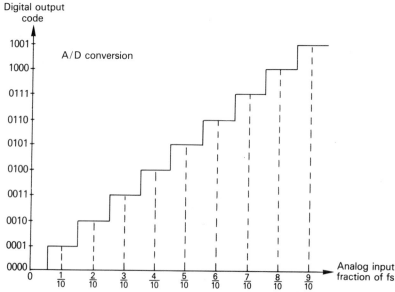

that is, the input range 7.5 to 6.5 V.

7. In the BCD code each digit in the decimal fraction requires a 4-bit binary word to represent it. Thus decimal 0.17 corresponds to BCD 0001 0111; decimal 0.56 corresponds to BCD 0101 0110; decimal 0.95 corresponds to BCD 1001 0101.

Answers to Self-assessment Exercises 227

8. (a) 0.3; (b) 0.5; (c) not valid; (d) 0.58; (e) not valid.

9. (a) − 1/4, + 10/16, − 2/16, + 1/32; (b) + 3/4, − 10/16, + 14/16, − 1/32; (c) + 3/4, − 6/16, + 14/16, − 31/32.

CHAPTER 2

1. (i) b; (ii) a; (iii) b, c; (iv) c; (v) a, d; (vi) b.

2. (a) an operational amplifier configured as a current to voltage converter. (b) the voltage range which can be applied to the output terminal without changing the value of the output current. (c) an external reference voltage which can be varied. (d) the input digital code and the variable reference voltage. (e) set the value of the reference current. (f) series negative feedback applied to the reference amplifier. (g) into the output terminal. (h) switch the polarity of the output. (i) complementing the MSB. (j) hold the analog value corresponding to the digital input data present at some instant in time. (k) a shift register and a storage register. (l) the time taken for the output to reach within $\pm \frac{1}{2}$ LSB of a new value following a change in digital input. (m) acquisition time.

3. 10 kΩ, 20 kΩ, 40 kΩ, 80 kΩ. (a) 1.875 mA; (b) 1.375 mA; (c) 0.375 mA.

4.

5.

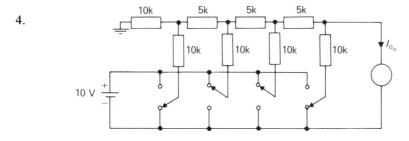

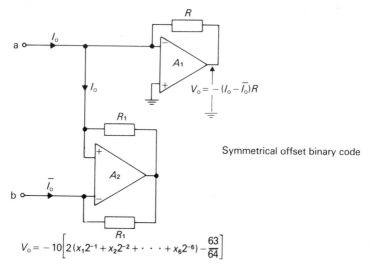

Symmetrical offset binary code

6. Digital input 10001000 gives: $I_o = (136/256) I_{ref} = 1.0625$ mA therefore $I_{ref} = 2$ mA and $\bar{I}_o = (119/256) I_{ref} = 0.9297$ mA. 01111111 gives: $I_o = (127/256) I_{ref} = 0.9922$ mA and $\bar{I}_o = (128/256) I_{ref} = 1$ mA. 01110111 gives: $I_o = (119/256) I_{ref} = 0.9297$ mA and $\bar{I}_o = (136/256) I_{ref} = 1.0625$ mA.

7. $I_{ref} = V_{ref}/R_{ref}$; with $V_{ref} = 10$ V, $I_{ref} = 2$ mA, then $R_{ref} = 10/2 = 5$ kΩ. Nominal full scale range (full scale negative to full scale positive) $= I_{ref} R = 20$ V, therefore $R = 20/2 = 10$ kΩ. Equation 2.11 gives

$$V_o = 20 [(x_1 2^{-1} + x_2 2^{-2} + \ldots + x_6 2^{-6}) - \tfrac{1}{2}]$$

(a) Code 000000 gives $V_o = -10$ V. (b) Code 000101 gives

$$V_o = 20 \left(\frac{4+1}{64} - \frac{32}{64} \right)$$

$$= -8.4375 \text{ V}$$

(c) Code 011111 gives

$$V_o = 20 \left(\frac{31}{64} - \frac{32}{64} \right)$$

$$= -0.3125 \text{ V}$$

(d) Code 100001 gives

$$V_o = 20 \left(\frac{33}{64} - \frac{32}{64} \right)$$

$$= +0.3125 \text{ V}$$

Answers to Self-assessment Exercises   229

(e) Code 111111 gives

$$V_o = 20 \left(\frac{63}{64} - \frac{32}{64} \right)$$

$$= +9.6875 \text{ V}$$

8. $I_{ref} = V_{ref}/R_{ref} = 12/(3.3 + 2.7) = 2 \text{ mA}$

$$I_o = I_{ref} \left(x_1 2^{-1} + x_2 2^{-2} + \ldots + x_8 2^{-8} \right) \quad \bar{I}_o = \frac{255}{256} I_{ref} - I_o$$

$$V_o = \left(I_o - \bar{I}_o \right) R_2$$

(a) Code 00000000 gives $I_o = 0; \bar{I}_o = (255/256) I_{ref} = 1.9922 \text{ mA}$

$$V_o = - \frac{255}{256} \times 2 \times 3.9 = -7.7695 \text{ V}$$

(b) Code 00000011 gives $I_o = (3/256) I_{ref} = 0.0234 \text{ mA}; \bar{I}_o = (252/256)$
$I_{ref} = 1.9687 \text{ mA}$

$$V_o = - \frac{249}{256} \times 2 \times 3.9 = -7.5867 \text{ V}$$

(c) Code 10000000 gives $I_o = (128/256) I_{ref} = 1 \text{ mA}; \bar{I}_o = (127/256)$
$I_{ref} = 0.9922 \text{ mA}$

$$V_o = + \frac{1}{256} \times 2 \times 3.9 = 0.0305 \text{ V}$$

(d) Code 10000001 gives $I_o = (129/256) I_{ref} = 1.0078 \text{ mA}; \bar{I}_o = (126/256)$
$I_{ref} = 0.9844 \text{ mA}$

$$V_o = + \frac{3}{256} \times 2 \times 3.9 = +0.0914 \text{ V}$$

(e) Code 11111110 gives $I_o = (254/256) I_{ref} = 1.9844 \text{ mA}; \bar{I}_o = (1/256)$
$I_{ref} = 0.0078 \text{ mA}$

$$V_o = \frac{253}{256} \times 2 \times 3.9 = 7.7086 \text{ V}$$

9.

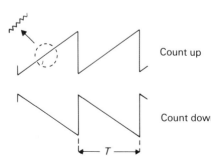

9. $$T = 2^{10} \times T_{cp} = 2^{10} \times \frac{1}{10^6}$$

$$f = \frac{1}{T} = \frac{10^6}{2^{10}} = 977 \text{ Hz}$$

10. Time to acquire new output value = $T_s + T_a = 2.5 \text{ }\mu\text{s}$

$$f_{max} = \frac{1}{2.5 \times 10^{-6}} = 4 \times 10^5 \text{ Hz}$$

CHAPTER 3

1. (a) Comparator, clock source, digital counter, DAC; (b) comparator, clock source, successive approximation register, DAC; (c) comparator, logic encoder; (d) comparator, integrator, clock source, digital counter.

2. Successive approximation A/D converter, parallel A/D converter, quantised feedback A/D converter.

3. (a); (c) i; (d) i, iii; (f).

4. 63

5. (a) serial to parallel. (b) make valid digital data continuously available. (c) the dual-slope technique. (d) the parallel conversion technique. (e) reference.

6. (a) Full scale code is 11111111. Converter requires 255 clock pulses. Conversion time = $255 \times T_{cp} = 255 \,(1/2 \times 10^{-6}) = 127.5 \text{ }\mu\text{s}$.
(b) Nominal full scale input is $I_{ref}R_{in} = (V_{ref}/R_{ref})\,R_{in} = 5$ V. 4.5 V is $(4.5 \times 256)/(5 \times 256) = 230.4/256$ of nominal full scale. Counter must increment up 231 counts. $T_{conv} = 231 \times 0.5 \text{ }\mu\text{s} = 115.5 \text{ }\mu\text{s}$. Digital code is 11100111.

7. $R_{ref} = V_{ref}/I_{ref} = 10/2 = 5 \text{ k}\Omega$. For ± 10 V range require $I_{ref}R_{in} = 20; R_{in} = 10 \text{ k}\Omega$.

$$\text{Loop slew rate} = \frac{I_{ref}R_{in}}{2^n} \times f_{cp}$$

$$= \frac{20}{64} \times 10^6 = 0.31 \text{ V}/\mu\text{s}$$

Answers to Self-assessment Exercises 231

$2\pi f_{max} \times 5 <$ loop slew rate; $f_{max} < 9.947$ kHz.

8. *Unknown number 11*: 32 high, 16 high, 8 low, 12 high, 10 low, 11 correct.
Unknown number 53: 32 low, 48 low, 56 high, 52 low, 54 high, 53 correct.

9. $I_{ref} R_{in} = 2 \times 5 = 10$ V; 4.5 V is $(4.5 \times 256)/(10 \times 256) = (115.2/256)$ of full scale. Values of I_o/I_{ref} produced in the conversion sequence are

$\dfrac{I_o}{I_{ref}}$	$\dfrac{127}{256}$	$\dfrac{63}{256}$	$\dfrac{95}{256}$	$\dfrac{111}{256}$	$\dfrac{119}{256}$	$\dfrac{115}{256}$	$\dfrac{117}{256}$	$\dfrac{116}{256}$
Comparator output	0	0	1	1	0	1	0	0

The conversion produces the digital code 01110100 corresponding to

$$V_{in} = \left(\frac{116}{256} \pm \frac{0.1}{256}\right) I_{ref} R_{in}$$

10. $I_{ref} = V_{ref}/R_{ref} = 2$ mA; full scale range $(\pm)$, $I_{ref}R_{in} = 20$ V. 3 V is $3/20 \times 256/256 = 38.4/256$ of fs range. In conversion sequence if

$$-\frac{38.4}{256} > \frac{\bar{I}_o}{I_{ref}} - \frac{128.5}{256} \quad \text{comparator gives '0'}$$

if

$$-\frac{38.4}{256} < \frac{\bar{I}_o}{I_{ref}} - \frac{128.5}{256} \quad \text{comparator gives '1'}$$

$\dfrac{\bar{I}_o}{I_{ref}}$	$\dfrac{128}{256}$	$\dfrac{64}{256}$	$\dfrac{96}{256}$	$\dfrac{80}{256}$	$\dfrac{88}{256}$	$\dfrac{92}{256}$	$\dfrac{90}{256}$	$\dfrac{91}{256}$
$\left(\dfrac{\bar{I}_o}{I_{ref}} - \dfrac{128.5}{256}\right)$	$\dfrac{-0.5}{256}$	$\dfrac{-64.5}{256}$	$\dfrac{-32.5}{256}$	$\dfrac{-48.5}{256}$	$\dfrac{-40.5}{256}$	$\dfrac{-36.5}{256}$	$\dfrac{-38.5}{256}$	$\dfrac{-37.5}{256}$
Comparator	1	0	1	0	0	1	0	1

that is, 3 V produce the code 10100101 (x_i values). This satisfies equation 3.10, that is

$$\frac{-37.9}{256} > \left(\bar{x}_1 2^{-1} + \bar{x}_2 2^{-2} + \ldots + \bar{x}_8 2^{-8}\right) - \tfrac{1}{2} > -\frac{38.9}{256}$$

Note $\left(\bar{x}_1 2^{-1} + \bar{x}_2 2^{-2} + \ldots + \bar{x}_8 2^{-8}\right) - \tfrac{1}{2} = \dfrac{38}{256}$

Full scale positive code 00000000 for

$$V_{in} = \left(\frac{127}{256} \pm \frac{0.5}{256}\right) 20$$

$V_{in} = 9.92 \text{ V} \pm 0.04 \text{ V}$

Zero scale code 01111111 for

$$V_{in} \pm \frac{0.5}{256} \times 20 = 0 \pm 0.04 \text{ V}$$

Full scale negative code 11111111 for

$$V_{in} = \left[-\frac{128}{256} \pm \frac{0.5}{256}\right] 20$$

$V_{in} = -10 \pm 0.04 \text{ V}$

11. Conversion time $= (n + 1) T_{cp} = 11 (1/2 \times 10^6) = 5.5 \ \mu s$.

12. $N_x = N_i V_{in}/V_{ref}$ (equation 3.12)

$$V_{ref} = \frac{1000}{1999} \times 1.999 = 1 \text{ V}$$

(a) Highest clock frequency which gives max rejection of 50 Hz pick up is that for which $T_i = N_i T_{cp} = 1/50$, that is

$$f_{cp} = \frac{1}{T_{cp}} = 50 \times N_i = 50 \text{ kHz}$$

(b) Similarly for 60 Hz rejection require

$$f_{cp} = 60 \times N_i = 60 \text{ kHz}$$

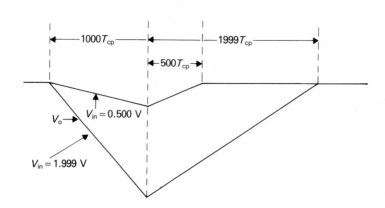

Answers to Self-assessment Exercises 233

For f_{cp} = 50 kHz, V_{in} = 1.999 V gives $V_o = (V_{in}/CR) 1000T_{cp} \triangleq$ 4 V; V_{in} = 0.500 V gives V_o = 1 V. For f_{cp} = 60 kHz, V_{in} = 1.999 V gives V_o = 3.33 V, V_{in} = 0.500 V gives V_o = 0.83 V. For 19.99 V full scale use V_{ref} = 10 V and increase integrator CR time constant say 10 times so that integrator output does not exceed 4 V.

13. Figure 3.10 with V_{ref} = 10 V, R_{ref} = 5 kΩ, R_{in} = 10 kΩ and resistor $128R_{ref}$ used to set $\pm \frac{1}{2}$ LSB quantisation uncertainty range. Use SAR bit 7 register to indicate end of conversion (see figure 3.11).

14. (a) α = + 0.2; + 0.2 = + 51.2/256. Gives digital code (x_1s) 10110010 for mid-range value 51/256.
(b) α = − 0.1; − 0.1 = 25.6/256. Gives digital code 01100101 for mid-range value − 26/256.

CHAPTER 4

1. Bit 1 switch wafer: switch points 0, 1, 2, 3, 4, 5, 6, 7 connected to ground, switch points 8, 9 connected to + 5 V. Bit 2 switch wafer: switch points 0, 1, 2, 3, 8, 9 connected to ground, switch points 4, 5, 6, 7 connected to +5 V. Bit 3 switch wafer; switch points 0, 1, 4, 5, 8, 9 connected to ground, switch points 2, 3, 6, 7 connected to + 5 V. Bit 4 switch wafer: switch point 0, 2, 4, 6, 8 connected to ground, switch points 1, 3, 5, 7, 9 connected to + 5 V. I_{ref} = V_{ref}/R_{ref} = 2 mA; value required for R_i = 5 kΩ.

2. Operational amplifier configured as in figure 4.3, with say R_1 = 200 Ω R_2 = 1 kΩ makes $I_L = I_o (R_2/R_1)$ = $5I_o$. Output voltage compliance limit $\triangleq 12 - I_L R_1$ = 10 V. Maximum value of load resistor = 1 kΩ.

3. (a) $\dfrac{V_o}{V_{in}} = \dfrac{R_2}{R_1} \left(\dfrac{32 + 4}{64} \right) = 1.13$

(b) $\dfrac{V_o}{V_{in}} = \dfrac{R_2}{R_1} \left(\dfrac{16 + 8 + 2 + 1}{64} \right) = 0.84$

I_{ref} must flow into DAC 1, therefore

$\left| \dfrac{V_{in}}{-ve} \right| < V_s \dfrac{R_1}{R_{ref}} < 10 \text{ V}$

4. (a) 0100000000; (b) 0000100000; (c) 0000000100.

5. (a) DAC's output current flows as collector current of Q_1; capacitor charging

current flows as collector current of Q_2 towards integrator summing point, integrator output ramps down.
(b) DAC's output current flows through D_1 and constitutes the capacitor charging current which flows away from the integrator summing point, integrator output ramps up.

6. (a) Charge and discharge currents are equal to DAC's output current I_o = $(15 \times 1)/(6.8 \times 4)$ = 0.55 mA.
(b) Slope = I_o/C = 5.5×10^4 V/s
(c) Peak-to-peak value of triangular wave = $20 \times (2.2/4.7)$ = 9.4 V
(d) $f = 4.7/4 \times 10 \times 2.2 \times 10^{-8} \times 0.55 \times 10^{-3}$ = 2.9 kHz

7. (a) Capacitor current supplied by DAC causes integrator output to ramp up. Current is I_o = 0.55 mA. Capacitor current supplied as Q_2 collector current causes integrator output to ramp down. Current is 0.55/2 mA.
(b) Positive slope = I_o/C = $+ 5.5 \times 10^4$ V/s. Negative slope = $I_o/2C$ = $- 5.5 \times 10^4$ V/s.
(c) As in 6c.
(d) Ramp up time $T_1 = 9.4/(5.5 \times 10^4)$ s. Ramp down time $T_2 = 2 \times 9.4/(5.5 \times 10^4)$ s

$$f = \frac{1}{T_1 + T_2} = \frac{5.5 \times 10^4}{3 \times 9.4} = 1.96 \text{ kHz}$$

8. *A/D converter*

$$V_{in} = S_1 \left(x_1 2^{-1} + x_2 2^{-2} + x_3 2^{-3} + x_4 2^{-4} \right) \pm \frac{1}{32} S_1$$

For V_{in} = 10 V $\pm \frac{1}{2}$ LSB to give code 1111 we require $S_1 = 10 \times 16/15$ = 10.67 V.

D/A converter

$$V_o = S_2 \left(x_1 2^{-1} + x_2 2^{-2} + \ldots + x_8 2^{-8} \right)$$

For full scale digital code to produce 10 V out we require $S_2 = 10 \times 256/255$ = 10.04 V.
Input scaling: let $15/16 \, S_1$ = 10 V represent 90° that is LSB represent 6°.
Output scaling: let V_o = 10 V represent sin 90° = 1 produced by digital code 11111111 applied to the DAC. Must programme ROM's memory location 15 with 11111111. Other values shown in the table opposite.

9. (a) A single-quadrant sine variation continuously repeating ($\frac{1}{4}$ of a sine wave). (b) A two-quadrant sine variation (positive half of a sine wave). Modification required is a swiching of the DAC's output polarity every time count mode changes from down to up.

Answers to Self-assessment Exercises 235

Sinusoidal frequency = $\dfrac{f_{cp}}{4 \times 16}$

10. (a) A repetitive bipolar triangular wave with limits + 10 V and − 10 V

frequency = $\dfrac{10^6}{2 \times 256}$

(b) A repetitive unipolar triangular wave with limits 0 V and − 10 V

frequency = $\dfrac{10^6}{2 \times 256}$

(c) A repetitive positive going ramp between the limits 0 V and + 10 V

frequency = $\dfrac{10^6}{256}$

Table for question 8

θ ($\pm 3°$)	V_{in} $\times S_1 \pm \tfrac{1}{32} S_1$	N = 255 sin θ (nearest whole number)	V_o = $S_2 \times \tfrac{N}{256}$	Memory address (DAC output)	Memory location	Data stored in memory locations
0	0	0	0	0 0 0 0	0	0 0 0 0 0 0 0 0
6	1/16	27	1.06	0 0 0 1	1	0 0 0 1 1 0 1 1
12	2/16	53	2.08	0 0 1 0	2	0 0 1 1 0 1 0 1
18	3/16	79	3.10	0 0 1 1	3	0 1 0 0 1 1 1 1
24	4/16	104	4.08	0 1 0 0	4	0 1 1 0 1 0 0 0
30	5/16	127	4.98	0 1 0 1	5	0 1 1 1 1 1 1 1
36	6/16	150	5.88	0 1 1 0	6	1 0 0 1 0 1 1 0
42	7/16	171	6.71	0 1 1 1	7	1 0 1 0 1 0 1 1
48	8/16	190	7.45	1 0 0 0	8	1 0 1 1 1 1 1 0
54	9/16	206	8.08	1 0 0 1	9	1 1 0 0 1 1 1 0
60	10/16	221	8.67	1 0 1 0	10	1 1 0 1 1 1 0 1
66	11/16	233	9.14	1 0 1 1	11	1 1 1 0 0 1 0 1
72	12/16	243	9.53	1 1 0 0	12	1 1 1 1 0 0 1 1
78	13/16	249	9.76	1 1 0 1	13	1 1 1 1 1 0 0 1
84	14/16	254	9.96	1 1 1 0	14	1 1 1 1 1 1 1 0
90	15/16	255	10.00	1 1 1 1	15	1 1 1 1 1 1 1 1

CHAPTER 5

1. (a) switch several input signals one at a time. (b) process analog signals and convert them into digital form. (c) quantisation uncertainty. (d) a switch and a capacitor. (e) equal to its input signal. (f) equal to the value of the input signal at the instant the hold command was applied. (g) the time taken for its output to become equal to its input (within a specific error) when the sample/hold is switched from hold to sample. (h) no analog sample/hold module is required. (i) input signals are slowly varying. (j) greater than twice the frequency of the highest frequency component of the analog signal. (k) the first word written into the memory is the first word which appears at the memory output when data is recalled. (l) slew rate.

2. Converter, quantisation error, scale error, offset error, non-linearity error. Sample/hold, gain error, offset error. Scaling and offset errors in analog conditioning circuitry.

3. Number of channels. Speed. Accuracy. Cost

4. (a)

5. (b)

6. (b), (d)

7. (a) SNR = 313.5 = 49.93 dB (see equation 5.2)
   (b) SNR = 156.8 = 43.9 dB

8. $T_s = (8 + x) T_{cp}$; $T_{cp} = 0.5$ μs therefore with $T_{aq} = 4.2$ μs, $x = 9$ therefore $f_s = (2 \times 10^6/17) = 118$ kHz

9. $T_d = K/f_s = 256/10^5 = 2.56$ ms

10. $T_d = (K + 1)/f_s$ where $K = (512/8)$ therfore $T_d = 1.3$ ms

11. $f_s = 512/T_d = 512/(5 \times 10^{-3}) = 102.4$ kHz. Bandwidth limit set by sampling frequency limitation $< 102.4/2$ that is < 51.2 kHz

12. (a) Equation 5.9

$$\left| \frac{\Delta V_{in}}{\Delta_t} \right| = \frac{V_{fs}}{2^n T_{conv}}$$

$$= \frac{20}{2^{10} \times 12 \times 10^{-6}} = 1.628 \times 10^3 \text{ V/s}$$

Answers to Self-assessment Exercises

Maximum sampling frequency = $1/T_{conv}$ = $1/(12 \times 10^{-6})$ = 83 kHz. Analog signal frequency limit set by slew rate (equation 5.10)

$$f_{max} = \frac{1}{\pi 2^n T_{conv}} = 26 \text{ Hz}$$

(b) Equation 5.12

$$\left.\frac{\Delta V_{in}}{\Delta t}\right|_{max} = \frac{V_{fs}}{2^n T_{ap}}$$

$$= \frac{20}{2^{10} \times 10^{-7}} = 195.4 \times 10^3 \text{ V/s}$$

Maximum sampling frequency

$$= \frac{1}{T_{aq} + T_{conv}} = \frac{1}{16 \times 10^{-6}} = 62.5 \text{ kHz}$$

Analog frequency limit as set by slew rate: $f_{max} = 1/(\pi 2^n T_{ap}) = 3.11$ kHz.

Note that even using a sample/hold the finite aperture time of the sample/hold still means that a slew rate limitation determines the maximum frequency of a full scale sinusoidal signal that can be accurately digitised.

13. (a) Channel throughput rate

$$= \frac{1}{8(T_{mux} + T_{conv})} = \frac{1}{8 \times 15 \times 10^{-6}}$$

$$= 8.33 \times 10^3 \text{ samples/s/channel}$$

Slew rate limit sets maximum frequency of full scale sinusoidal signal as before at 26 Hz.

(b) Channel throughput rate =

$$= \frac{1}{8(T_{conv} + T_{ap} + T_{aq})} = \frac{1}{8 \times 16.1 \times 10^{-6}}$$

$$= 7.76 \times 10^3 \text{ samples/s/channel}$$

Sampling frequency limit would set bandwidth at $< 7.76 \times 10^3/2$ but full scale bandwidth limit as set by slew rate is < 3.11 kHz due to finite aperture time of sample/hold.

CHAPTER 6

1. See sections 6.2.1 and 6.2.2.

2. (a), (d), (f)

3. (a) 139/256 (b) 96/100 (c) + 11/25 (d) − 106/256 (e) + 22/128

4. (a), (c)

5. A ready built system which conditions, multiplexes and digitises analog input signals into a computer or microcomputer system. (b) it corresponds to the normal calibration procedure. (c) errors in the weighting of the DAC's bit current increments. (d) no greater than $\pm \frac{1}{2}$ LSB. (e) there are only $2^n - 1$ steps separating the 2^n analog states of an n-bit converter.

6. To ensure monotonicity, differential linearity error must be no greater than ± 1 LSB. Therefore differential linearity error drift over temperature must be no more than $\pm \frac{1}{2}$ LSB. This gives allowable temperature change as $\pm \theta$ °C where

$$4 \times 10^{-6} \times \theta < \frac{1}{2} \times \frac{1}{2^{12}}$$

that is

$$\theta < \frac{10^6}{8 \times 10^{12}}; \; \theta < 30\,°C$$

7. MSB = 2^9 LSB. Error of − 0.2 per cent in MSB represents an error of − (0.2/100) × 2^9 = − 1.02 LSB. Differential linearity error is − 1.02 LSB. At code transition 0111111111 to 10000000000 DAC output will decrease by 0.02 LSB instead of the correct 1 LSB increase. The transfer function will be non-monotonic.

8. Error budget:

Specification	% error
Quantisation uncertainty $\pm \frac{1}{2}$ LSB, $\frac{1}{2} \times 2^{-12} \times 100$	= 0.0120
Differential linearity error $\pm \frac{1}{2}$ LSB	= 0.0120
Differential linearity error drift over temperature $3 \times 10^{-6} \times 25 \times 100$	= 0.0075
Gain error drift over temperature $25 \times 10^{-6} \times 25 \times 100$	= 0.0625
Zero error drift over temperature $5 \times 10^{-6} \times 25 \times 100$	= 0.0125
Error due to power supply change 0.002×1	= 0.0020
Worst case error	± 0.1085%

Index

Accuracy *see* Error
  of sampled data system 160, 161
Acquistion time *see* Sample/hold
Addition of digital variables 100, 101
Aliasing 123
Analog input/output systems 181
Analog signals 1, 165, 121
  delay in 147-50
  digitisation of 121, 125-7
Analog to digital conversion techniques,
  comparison of 77-9
  dual-slope 63-7
  feedback 40, 41
  integrating 62-71
  parallel 72, 73
  quantised feedback 67-71
  ramp type 41-3
  ratiometric 73-7
  successive approximation 49-61
  tracking 43-9
Arithmetic operations 92-106
Attenuator, digitally programmed 89
Audio signals, delay in 147-50
  digital transmission of 143, 145, 146
  digitisation of 127
Auto polarity 65
Auto-zero 65, 189

Binary, code 3
  coded decimal (BCD) 8, 183
  counter 24, 115
  fractional code 3, 6, 183
  weighted resistor quads 20
  weighted resistors 14, 15
Bipolar, coded DACs 26-9
  codes 8-10, 183

successive approximation
  converter 58, 59
  tracking converter 44
Bit 2
  LSB 2
  MSB 2
  weights 3
Bus lines 166

Calibration procedures 201, 202, 204, 206, 208, 210, 217
Capacitive coupling 215
Circuit layout 60, 212-15
Codes 2, 183
  bipolar 8-10, 183
  complementary binary 57
  decimal 6-8
  fractional 3
  Gray 6, 7
  natural binary 3
  offset binary 9, 183
  sign magnitude 9, 184
  symmetrical offset binary 28
  two's complement 10, 32, 183
Companding DAC *see* Logarithmic DAC
Compression *see* Logarithmic compression
Conversion relationships 4-6, 182, 183
Conversion time, of ramp type converter 41, 43
  of successive approximation converter 59, 60, 162
  shortening 60
Converters, errors in *see* Error
  performance specifications of 181-98

239

Index

Converters (*cont.*)
  selecting 176, 179-81
Counter 24, 41, 42
Current, booster 83, 84
  inverter 29, 85, 97
  mirror 108
  source 84-7
  switching 19, 20
  to voltage conversion 29, 88, 89, 93

DAC, analog readout 15, 16
  bipolar 26-9
  four quadrant multiplying 93-100
  logarithmic 140-3
  offset binary coded 26
  sign magnitude coded 28-30
  symmetrical offset binary coded 27, 28
  two quadrant multiplying 88, 90, 93, 94
  two's complement coded 30, 31
Data acquisition systems 152-64
  accuracy of 160, 161
  analog multiplexed 154-7
  characterisation of 158, 160
  speed of 162-4
  with a converter per channel 158, 159
Data bus 166, 167
Decimal codes 6-8
Delay line 147-50
Design procedures 177-9
Differential input measurement amplifier 75
Digital control of, frequency 107-11
  gain 87-92
  ramp rate 112-14
  timing period 113-14
  waveform generator 108-11
Digital generation of analog waveforms 115-17
  generation of functional relationships 114, 115
Digital memory 114, 115
Digital multiplication 102-5
Digital offset nulling 107, 108
Digital scale setting 106, 107
Digitisation of analog signals 125-40
Distortion due to quantisation error 125
Dual-slope A to D converter 63-7

Error, due to bit current weighting 192
  due to temperature change 196-7
  extraneous sources of 212-17
Error analysis 199-212
Error budget 211
Error evaluation examples 200-11

Feedback A to D converters 40, 41
Feedback loop 43
Filter, low pass 43
Flash encoder *see* Parallel A to D converter
Frequency compensation of reference amplifier 24, 95
Functional operations 82, 114, 115

Gain control 87-92
Glitches 33-6
  deglitcher 36
Gray code 6, 8
Grounding 60, 213-16

Integrating A to D converters 62-71
Integrator 63, 108
Interference noise 212-13

Latch *see* Storage register
Layout *see* Circuit layout
Log amplifier 137
Logarithmic compression 134, 137
Logarithmic DAC 140-3
Linearity specifications 191-6
  differential linearity error 192
  integral linearity error 192

Major code change 35, 195
Measurement amplifier 75, 154, 215
Memories, RAM 150, 152
  ROM 114, 115
  sequentially accessed 145, 147, 150
Memory mapped input/output 170
Microprocessor interface with converters 164-72
Missing code 195, 196
Monotonicity 192
Multiplexing 154, 162
Multiplication of digital variables 102-5
Multiplying DAC 20
  differential input 97, 98
  four quadrant 93-100
  two quadrant 88, 90, 93, 94

Index

Noise, due to quantisation error 125-7
  rejection 62, 63
  SNR 127
Non-linear encoding 134-43, 145
Nybble 168

Offset binary code 9, 183
Offset binary coded DAC 26, 27
Offset error 161, 187-9, 203-10
Operational amplifier, as current to voltage converter 15, 16
  as high input impedance follower 15, 16
  current source 83-7
  used to set DAC reference current 21
Output polarity switching 30

Parallel A to D converter 72, 73
Parallel data transmission 2
Performance specifications 181-98

Quantisation, distortion 125
  error 125
  noise 125
  non-linear 128, 134-43
  of analog range 4, 6, 181
  uncertainty 6, 43, 184-6
Quantised feedback A to D converter 67-71

Ramp 112
Ratiometric A to D conversion 73-7
Reference amplifier 21, 22, 24, 95
Reference connections 21, 22, 24
Registers 32, 33, 61, 62, 68, 147, 167
Resistance measurement 75-7
Resistive bridges 74, 75, 77
Resistor networks, binary weighted 14, 15
  R-2R ladder 17-19
Resolution 184-6
ROM 114, 115

Sample/hold 36, 46, 128, 132, 156, 163, 164
  acquisition time 36, 132, 164
  aperture time 163
  deglitcher 36
  function performed by tracking converter 45, 46, 130, 131
Sampled data systems 128-71

applications of 143-52
  practical system 132-4
Sampling 125
  theorem 121-3, 125, 160
Scale factor 182, 189
Serial data transmission 2, 33, 171, 172
Serial to parallel data conversion 33, 34
  performed by SAR 50, 51
  performed by shift register 33, 34, 147
Settling time 33-5
Shift register 33, 145, 147-50
Sign magnitude, code 9, 184
  coded DAC 28-30
Signal to noise ratio (SNR) 127
Skipped code 195, 196
Slew rate 36, 160, 162-4
  of tracking converter 45, 46, 130, 131
Speed limitations 77-9, 162-4
Storage register 32, 33, 61, 62, 147
Strobe 33
Subtraction of digital variables 100, 101
Successive approximation A to D converter 49, 50, 52-61
  bipolar 58, 59
  conversion sequence 54, 55
  conversion time 59, 60
  in sampled data system 130
  with output storage register 61, 62
Successive approximation register (SAR) 50, 51
Switching bit currents 14-20
Switching polarity of DAC output 29-30
Symmetrical offset binary code 27-9, 183

Temperature dependence 196, 199, 200
Thermal effects 215, 217
Throughput rate 158, 160
Timer 110
Timing capacitor 110
Timing diagram for SAR 51
Tracking A to D converter 43-9, 76, 106
  in sampled data system 130
Transducer 154, 156, 165

Index

Transient errors 34–6
Transient recording 147, 150–3
Truncating SAR 60, 61, 134
Two's complement code 10, 59, 183

Voltage compliance 24, 28, 85, 110

Voltage source 82, 83

Waveform generation 114–17
Waveform generator, digitally controlled 108–11